Chemistry in the Garden

Chemistry in the Garden

James R. Hanson
Department of Chemistry, University of Sussex, Brighton, UK

RSC Publishing

ISBN: 978-0-85404-897-7

A catalogue record for this book is available from the British Library

Published by The Royal Society of Chemistry,
Thomas Graham House, Science Park, Milton Road,
Cambridge CB4 0WF, UK

Registered Charity Number 207890

For further information see our web site at www.rsc.org

Printed and bound in Great Britain by MPG Books Ltd, Bodmin, Cornwall

Foreword

I am very pleased to have been asked to provide a foreword for this authoritative account of the chemical interactions and linkages between garden plants and practices and chemistry, as these subjects are rarely referred to in any detail, even in books devoted to the science of gardening.

While opinions, often entrenched, are expressed on the use or misuse of chemical sprays for controlling weeds, pests and diseases and the effects on plants of the application of inorganic or organic fertilisers, seldom is any consideration given by gardeners, whether professional or amateur, to the ways in which chemistry is so intimately connected with their own hobby or profession and its importance in understanding the plants they grow and their cultivation.

As Dr Hanson indicates in his Preface, chemistry and chemical ecology play a central role in determining a great deal of what happens in gardens and the numerous chemical interactions that take place between the soil and the plants we grow. As a non-chemist with only distant memories of the structures of amino acids, polysaccharides, polyketides and many other chemical compounds mentioned here and once studied in relation to my botanical background, I read this book with a degree of trepidation. I need not have had concerns, however, as it is written in a very informative style that follows a logical pattern throughout with, subject by subject, linkages between the chemistry and the elements of gardening being clearly established.

Dr Hanson has concentrated primarily on the most interesting chemical and ecological compounds that are important in terms of the cultivation of vegetables, fruit and ornamental plants, the soil in which they grow and the natural interactions between plants and the insect pests and disease organisms that attack them. Throughout this book there is a wealth of information on the natural chemical background upon which so many garden plants depend both to survive and thrive. These include natural defence mechanisms – anti-feedants within plants – to help combat insect pests and diseases; root exudates as in lily-of-the-valley to deter the spread of competing plants; the chemical background to colour and scent in garden plants and to the medicinal properties attributable to the cardiac glycosides in foxgloves; and the anti-bacterial and anti-fungal activity of allicin derived from garlic. These are just a few of the wide-ranging subjects covered in this stimulating account of chemistry in the garden.

While this book is primarily aimed at fellow chemists who are keen gardeners, it provides much factual and very useful information that will also be of considerable interest and value to professional and amateur gardeners as well as degree level students in the natural sciences who I trust will enjoy and benefit from its contents as I have.

Chris Brickell CBE, VMH
Former Director RHS Garden, Wisley
and former Director General, RHS

Preface

Chemistry determines much of what happens in the garden. The growth, the colour and the scent of plants and many of the interactions between species in the garden have a chemical basis. The aim of this book is to describe some aspects of the chemistry and chemical ecology which are found in the garden. The natural products that are described in the major undergraduate textbooks were often isolated from medicinal plants many of which came from the tropics. With a few exceptions, common garden plants, despite their availability and diversity, rarely figured in this discussion. In the garden there are numerous interactions between plants, the soil and with other organisms in which chemistry plays a central mediating role.

Many chemists are amateur gardeners and this book is directed at them. The book arose from discussions over coffee in which colleagues proudly described their triumphs at the local horticultural shows and I, as a bystander, was left wondering what were the structures of the compounds and the chemistry in the garden that had contributed to their successes.

The Royal Horticultural Society's *Good Plant Guide* lists some 3000 recommended plants for British gardens. Many of the common plants have several cultivars differing in size, colour and disease resistance, all factors in which chemistry plays a role. Furthermore, there is a seasonal variation in the formation of some natural products, whilst others are produced as a consequence of insect or fungal attack. There are considerable regional variations in the soil and local climate and hence in the plants that can successfully be grown in the garden. What may be present in a plant grown in one garden may not necessarily be present in another cultivar grown elsewhere. Consequently I have restricted discussion to a limited number of the chemically and ecologically interesting compounds that are produced by common ornamental garden plants, fruit and common vegetables and by the predators that attack them. Nevertheless I hope that this short book stimulates interest in the chemistry of the garden.

As with chemical nomenclature, botanical nomenclature is undergoing a continuous process of change. I have tried, with the helpful advice of my colleague, David Streeter of the School of Life Sciences in the University of Sussex, to use the currently accepted names for plants and their families. I wish to thank Dr Merlin Fox of the Royal Society of Chemistry, Dr Christopher Brickell and Dr Alex Nichols for their constructive and helpful comments in the preparation of this book.

Dr J. R. Hanson
University of Sussex

Contents

Chapter 1 Introduction

1.1 Chemical Diversity in Plants 2
1.2 The Structure Elucidation of Natural Products 3
1.3 The Ecological Role of Natural Products 4
1.4 Changes in the Garden 6

Chapter 2 The Biosynthetic Relationship of Natural Products

2.1 Polyketides 12
2.2 Terpenoids 13
2.3 Phenylpropanoids 15
2.4 Alkaloids 17

Chapter 3 Natural Products and Plant Biochemistry in the Garden

3.1 The Structural Material of Plants 21
3.2 Photosynthesis 26
3.3 Oxidative Coenzymes 29
3.4 Plant Hormones 30

Chapter 4 Garden Soils

4.1 The Mineral Structure of the Soil 35
4.2 The Organic Content of the Soil 37
4.3 Nutrients from the Soil 38
4.4 The Role of pH 39
4.5 Fertilizers and Compost 40
4.6 Microbial Interactions within the Soil 41

Chapter 5 The Colour and Scent of Garden Plants

5.1 Colouring Matters 44
5.2 The Carotenoids 44

5.3 The Anthocyanins 46
5.4 Natural Pigments 50
5.5 Floral and Leaf Scents 52

Chapter 6 Bioactive Compounds from Ornamental Plants

6.1 Compounds from the Lamiaceae 59
6.2 The Foxgloves and Cardiac Glycosides 62
6.3 Poppies 63
6.4 Compounds from the Asteraceae 64
6.5 The Constituents of Bulbs 67
6.6 Toxic Compounds from Ornamental Plants 69
6.7 Compounds from Ornamental Trees 72
6.8 Mistletoe 75
6.9 Conifers 75

Chapter 7 Natural Products in the Vegetable and Fruit Garden

7.1 Root Vegetables 80
7.2 Onions, Garlic and Asparagus 86
7.3 The Brassicas 89
7.4 Lettuce 91
7.5 The Legumes 93
7.6 Rhubarb 94
7.7 Tomatoes 94
7.8 Fruit Trees 96
7.9 Soft Fruit 98

Chapter 8 Fungal and Insect Chemistry in the Garden

8.1 Microbial Interactions 107
8.2 Lichens 109
8.3 Mycorrhizal and Endophytic Organisms 110
8.4 Interactions between Fungi 111
8.5 Insect Chemistry in the Garden 113

Epilogue 122

Further Reading 124

Glossary 128

Subject Index 134

CHAPTER 1

Introduction

A garden can be a source of pleasure. The juxtaposition of different colours and floral scents together with the opportunity of eating home grown vegetables combine to provide gardeners with considerable satisfaction. This book is about the chemistry that is found in the garden and the chemical ecology involved in the interactions between the plants, micro-organisms and insects that live there.

The chemistry of the garden begins with the chemistry of the soil. Next to climatic conditions, what is present in the soil can have the biggest effect on the garden. The structure and chemistry of the soil determines the availability of nutrients, water and air to the roots. The pH of the soil and the mobility of metal ions, such as iron, affects the plants that can be grown. Part of the chemistry of gardening lies in achieving an appropriate mineral balance in the soil.

Soil and climatic conditions can have a marked impact on the chemistry of plants and the natural products that they form. For example plants that grow under arid conditions can produce a resinous or waxy covering on the leaves to reduce their water loss. The mineral content of the soil can vary the colour of flowers such as the hydrangea. Plants have the ability to take up metal ions and this can be used in the bioremediation of 'brown field' sites. It is also a warning to the vegetable gardener who may unwittingly ingest toxic metals such as cadmium.

The very act of gardening changes the chemistry of the soil. Digging, together with the distribution of fertilizer and compost, not only changes the air, mineral, water and organic content of the soil but also affects the presence of many chemical messengers within the soil. The smell of newly dug soil is an indication of the release of volatile chemicals produced by soil micro-organisms. More subtle effects involve the dispersion of insect trail substances and the redistribution of germination inhibitors produced by plants.

1.1 CHEMICAL DIVERSITY IN PLANTS

The chemistry that occurs within a plant is complex and highly organized. The organic compounds that occur in plants fall into three big groups. Firstly there are those compounds that occur in all cells and play a central role in the metabolism and reproduction of the cell. These are known as the primary metabolites and include the common sugars, the amino acids that are constituents of proteins, and the nucleic acids. The second group are the high molecular weight polymeric materials such as cellulose, lignin and the various proteins which form the structural and enzymatic components of the cell. The third group of naturally occurring compounds are those of relatively low molecular weight which are characteristic of a limited range of species. These are the secondary metabolites and they include the polyketides, the phenylpropanoids, the alkaloids and the terpenoids from which the colouring matters, the scent of flowers and the flavours of foods are derived.

Whereas many of the primary metabolites exert their biological effects within the cell in which they are produced, the secondary metabolites often exert their biological effects on other cells or other species. The range of secondary metabolites that are formed is a characteristic of particular species. Furthermore, structurally quite different secondary metabolites may play a similar role, for example, in the defence mechanisms of plants in different species. Many of the natural products that are of interest to the chemist in the garden are secondary metabolites. However the division between primary and secondary metabolites, whilst useful, is not rigid. Acids derived from primary metabolism form esters with secondary metabolites, and amino acids that are constituents of proteins are also the progenitors of alkaloids.

The typical garden displays considerable biodiversity. Even the average lawn will contain several different grasses as well as, probably, clover, daisies and moss! The variety of natural products that are found as a consequence of this biodiversity is wide. This chemical diversity is not restricted to differences between species. There is considerable infra-species variation which is reflected, for example, in the many colours exhibited by a single species such as the antirrhinum. The inheritance of colour in sweet peas laid the foundations of plant genetics. Furthermore the chemistry changes as the plant develops. This is exemplified by the changes in flavour and colour of fruit as it ripens. There is even a diurnal variation in the formation of some natural products. Flowers that are pollinated by moths are more highly scented in the evening, whilst those that rely on bees for pollination have their peak production earlier in the day.

Seedsmen have bred many cultivars of garden plants to enhance their visual properties and their resistance to disease. The organoleptic properties of vegetables have been modified, for example, to produce varieties of lettuce which do not possess the bitter taste of some sesquiterpenoid lactones. It has been reported that there are over 50 000 varieties of rose that have been produced commercially over several hundred years. In chemical terms, their natural product content has been varied. Whilst this variation may be in the finer detail of the balance between similar compounds, it can mean that some varieties do not produce a particular compound while others may even produce different natural products.

Although it is obvious in terms of plant pigments, the location of many other natural products in the plant varies widely. The constituents of the roots often differ from those of the aerial parts of the plant. Many natural products are produced as a response to a competitive stress. The nature of this competition varies between the roots and the aerial parts of the plant and this has an effect on the natural products that are produced. For example some salvias produce diterpenoid quinones in the roots with antibiotic properties and different diterpenoids (clerodanes) in the leaves as insect anti-feedants. Those natural products that are found in the flowers, or even parts of the flower, may differ from those that are found in the leaves, the stem or the fruits.

1.2 THE STRUCTURE ELUCIDATION OF NATURAL PRODUCTS

The quantities of the natural products which are produced by plants are usually small when compared to the fresh or even the dry weight of the plant. A good recovery of a natural product might be 100 mg from 1 kg dry weight and in many cases the amounts that are isolated are much smaller. These low concentrations can be a reflection of the biological activity of many natural products. However the amounts of any one compound may vary by as much as fiftyfold when, for example, it is a stress metabolite involved in the protection of a plant against microbial or insect attack.

The small quantities of natural product that are available has meant that the real diversity of structures has only become apparent over the past forty to fifty years since chromatographic methods of separation have been refined and spectroscopic methods of structure elucidation have become widespread. It is worth noting that chromatography was first introduced a hundred years ago in the context of the separation of natural products. Today many structures are established purely on the basis of spectroscopic, particularly nuclear magnetic resonance (NMR), studies.

The application of spectroscopic methods to organic chemistry has often been explored in the context of natural product structure elucidation. One of the first applications of ultraviolet (UV) spectroscopy to organic chemistry was nearly one hundred years ago in a study on berberine, the yellow alkaloid found in barberry (*Berberis* sp.). Theories relating the wavelength of the absorption maxima in the ultraviolet spectrum to the length of conjugated systems arose from the studies on the polyene carotenoid pigments of carrots and tomatoes. A useful set of correlation tables between structure and the frequencies of carbonyl absorption in the infrared (IR) spectrum, published in 1951, was produced in the context of the study of fungal metabolites including some compounds produced by *Penicillium gladioli*, an organism that is a plant pathogen found on the corms of gladioli. Many of the early studies utilizing NMR spectroscopy in the late 1950s and early 1960s involved natural products including, for example, the gibberellin plant hormones. Much of our knowledge of insect chemistry could not have been obtained without the combination of gas chromatography linked to mass spectrometry. This combination has also revolutionized the analysis of floral scents and the detection of plant hormones, which also occur in minute amounts.

A present day spectroscopic strategy for establishing the structure of a natural product involves determining the molecular formula from the high resolution mass spectrum. The functional groups and the number of double bonds are identified from the IR and NMR spectra. Consequently the number of rings can be established. At this stage the class of natural product is often clear. Most ‘new’ natural products are relatives of known compounds. A number of structural fragments may be apparent from the ^{1}H NMR spectrum. Further connectivities may be established by two-dimensional NMR spectra and the structure is then pieced together. The relative stereochemistry of the natural product may be established by a combination of coupling constant measurements and nuclear Overhauser effect experiments. The absolute stereochemistry may be established by NMR or circular dichroism methods. The power of modern X-ray crystallography is such that, given a suitable crystal, a structure may often be obtained in a matter of hours.

1.3 THE ECOLOGICAL ROLE OF NATURAL PRODUCTS

Many natural products, including those from garden plants such as the foxglove, were originally investigated because of their medicinal, perfumery or culinary value to man. However during the latter part of the twentieth century considerable attention was paid to the ecological role of natural products in mediating interactions between species. The chemistry of individual plants, fungi and insects in the garden cannot be considered in isolation. The chemistry of a plant is the summation of its intrinsic properties and the consequences of its interaction with its environment. A plant will produce a number of natural products in response to environmental challenges. A plant is essentially stationary. When attacked it cannot flee but it must fight. The formation of some natural products is elicited as a response to microbial or herbivore attack. These natural products can have a defensive role and may be toxic to the predator. The production of insect anti-feedants in leaves can be seasonal. Their presence in leaves prior to flowering can protect the plant against predators early in the season. After flowering and later in the year, the anti-feedants may not be present and the leaves are damaged prior to leaf fall and their decay. The damaged tissue provides a route of entry for micro-organisms involved in autumnal decay.

Some plants may exert their dominance of an area by producing natural products which inhibit the germination of seeds and prevent the growth of other seedlings. This competitive effect is known as allelopathy. Relatively few weeds will grow around mint. The constituents of the mint which inhibit the germination of seeds can be smelt by crushing the soil from around the plant.

When a fungus attacks a plant, the fungus may produce phytotoxins which damage cellular mechanisms together with enzymes which digest the plant material. The plant responds to this attack by producing natural anti-fungal agents known as phytoalexins. Attack by eelworms and other insects can also elicit the formation of natural products. Some root exudates are stimulants for the development of mycorrhizal fungi which can play an important part in mobilizing nutrients for the plant. The hatching of species:specific eelworms may also be stimulated by root exudates. Plants may contain insect anti-feedants and other toxic principles which protect the plant against attack by herbivores. In a further utilization of this activity, some butterflies can then sequester these plant toxins and use them for protection against predators such as birds. Other butterflies can overcome and metabolize the anti-feedants produced by their host plant and then, like the cabbage white butterfly, use them as oviposition stimulants in order to lay their eggs in a relatively non-competitive environment. The volatile signals released by plants to affect these aspects of insect behaviour are known as kairomones. In a further tripartite relationship some of the chemicals released by a plant infested with a grub can attract parasitic wasps to the grubs.

Many aspects of insect behaviour are regulated by chemical stimuli. The natural products that are involved are known as semiochemicals. Pheromones are semiochemicals that are involved in infra-species communication. These may involve trail substances, aggregation pheromones, and alarm and defensive substances as well as sex attractants. The effect of many of these in the garden can be easily observed. As we shall see in subsequent chapters, chemical communication and chemical warfare in the garden would put the Conventions on Chemical Warfare and the Health and Safety at Work Acts in the shade.

It is often claimed that vegetables grown under 'organic' conditions taste better than those grown under conventional conditions in which the environmental challenges are kept in check by insecticides, fungicides and herbicides. Some of these differences may be attributed to the formation of defensive substances by the plant in response to environmental challenges. However it is worth bearing in mind that these defensive compounds are produced by the plant to deter predators by their toxicity. Man in this context is a large predator.

1.4 CHANGES IN THE GARDEN

The introduction of a plant into a garden can have an impact on other plants as a consequence of the natural products which it produces. For example the root exudates of lily-of-the-valley will affect the germination and survival of neighbouring plants.

Changes in the global distribution of plants as a result of gardening can have an impact on natural products that is wider than anticipated. Rhododendrons were introduced into this country and have escaped. The toxic natural products produced by their flowers have been collected along with nectar by bees and can now present an occasional problem in honey. The escape of the giant hogweed, originally an ornamental plant (!) and the phototoxic effects of its components on the skin is a further example of an unanticipated gardening problem. Another deep-rooted alien invader which is giving cause for concern is Japanese knotweed (*Fallopia japonica,* syn. *Polygonum cuspidatum*).

The global transport of foodstuffs has a similar impact. For example it is worth remembering that the Colorado beetle was not originally a pest of potatoes but of another member of the Solanaceae. The progression of the pest from its original host plant to potatoes in the mid-west of the USA and thence to the eastern seaboard before coming to Europe has been well documented. Although other members of this family produce anti-feedants that are active against the Colorado beetle, the potato does not naturally produce them and cannot prevent an infestation. The arrival of the harlequin ladybird is a current cause of concern. The environmental perturbations brought about by transport have an impact on the ecological role of natural products.

Climatic change may not only affect the range of plants that can be grown in the garden and the yields of vegetable crops but also the metabolite production of common plants. Compare the evocative smell of a walk through a Mediterranean pine grove with a similar walk in the present day United Kingdom and then consider the impact of the increased volatility of their terpene content as the ambient temperature rises. In this context it is worth remembering that many of the semio-chemicals that affect insect behaviour are substances with a volatility which defines their zone of influence. Climate change may affect not just the plants that can be grown but also the micro-organisms that flourish in the garden. Whilst *Penicillium* species are typical of temperate climates, *Aspergilli* are more often found in warmer situations. Their characteristic metabolites are different and consequently there may be different problems arising from mycotoxin production. Climate change will have an impact on the chemistry of the garden.

CHAPTER 2

The Biosynthetic Relationship of Natural Products

The structural variety of the natural products that are found in plants can be rationalized in terms of their biosynthesis. This unifying biosynthetic framework not only serves to relate natural products one to another, but it can also reflect the botanical relationship between plants. The existence of a particular biosynthetic pathway in a plant reflects the presence of particular enzymes which in turn reflects the genetic make up of the plant. Not surprisingly botanically related plants produce similar natural products. In this chapter I present a brief outline of some of the major biosynthetic pathways that are found in plants and by which the structures of compounds described in the later chapters may be rationalized.

At first sight the structural diversity of natural products may seem bewildering. However the majority of natural products belong to one of several large families of compounds which have particular structural characteristics arising from the way in which they are assembled in nature, *i.e.* from their biosynthesis. These classes of secondary metabolites are:

polyketides and fatty acids
terpenoids and steroids
phenylpropanoids
alkaloids
specialized amino acids
specialized carbohydrates

In addition there are a number of simple carboxylic acids such as citric acid and malic acid which occur in the pathways of primary metabolism such as the tricarboxylic acid cycle. These are formed in large amounts in particular plants where they play a wider role than just that of primary metabolites.

The fundamental building blocks that plants use to assemble natural products such as phosphoenol pyruvate and acetyl coenzyme A arise from the catabolism of sugars. A 'map' illustrating these relationships is shown in the scheme. Carbon dioxide from the atmosphere is incorporated by photosynthesis into sugars. The breakdown of fructose 1,6-diphosphate by glycolysis affords glyceraldehyde monophosphate and dihydroxyacetone monophosphate. These provide the source of pyruvate and thence acetate units from which many natural products are formed. Acetyl coenzyme A is also used in the tricarboxylic acid cycle from which citric acid and malic acid are derived.

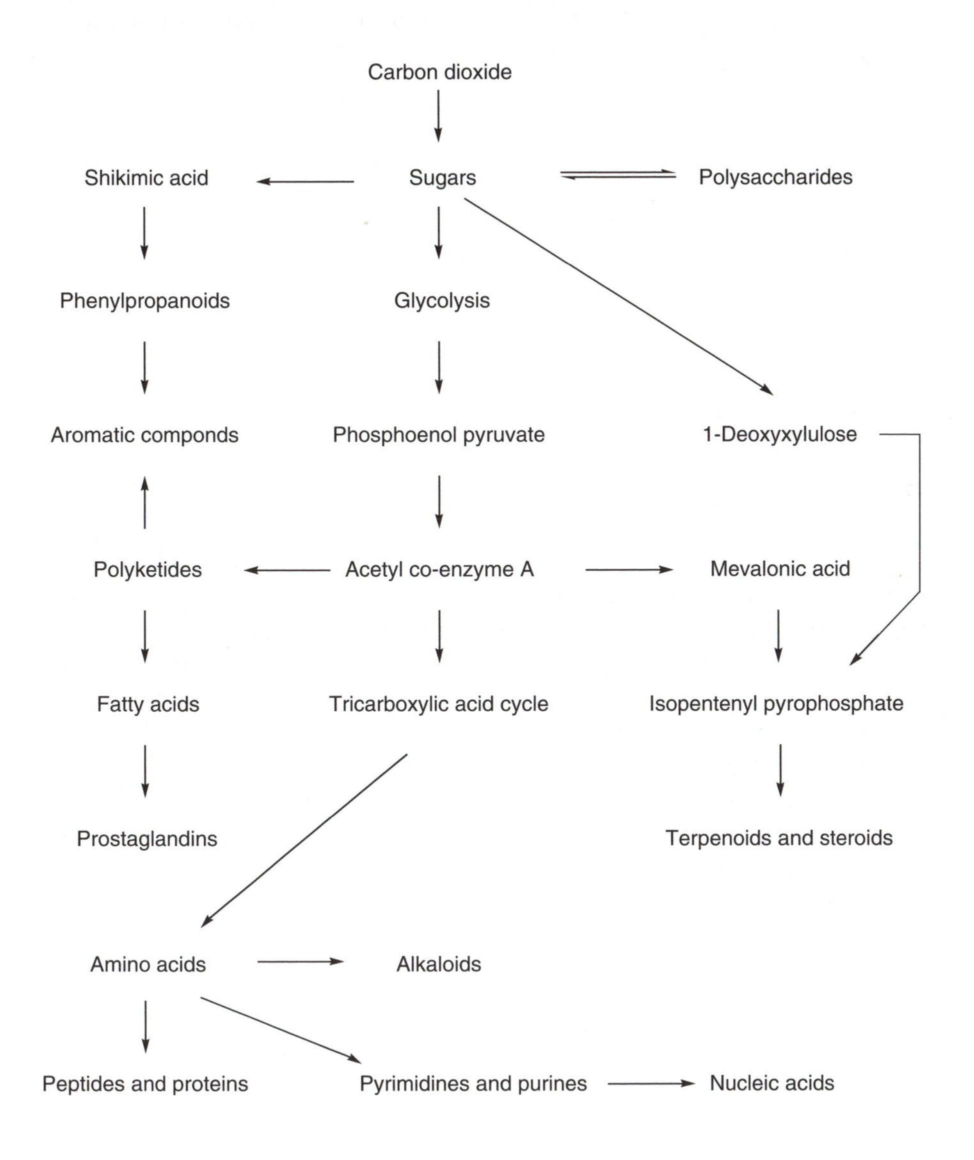

Polyketides and fatty acids are formed by the linear combination of acetate units derived from their building block, acetyl coenzyme A. An example is the unsaturated fatty acid, linoleic acid **2.1**, which occurs in many plant oils and is an essential dietary component for man. These compounds have a linear backbone with an even number of carbon atoms. Extra carbon atoms may be added from methionine. Terpenoids and steroids are assembled in nature from isoprenoid (C_5) units derived from isopentenyl (3-methylbut-3-en-1-yl) pyrophosphate **2.2**. These natural products have a characteristic $(C_5)_n$ branched chain structure and are exemplified by geraniol **2.3**, a component of rose oil. The phenylpropanoid group contain a C_6–C_3 unit. These compounds have at least one aromatic ring as in cinnamic acid **2.4**.

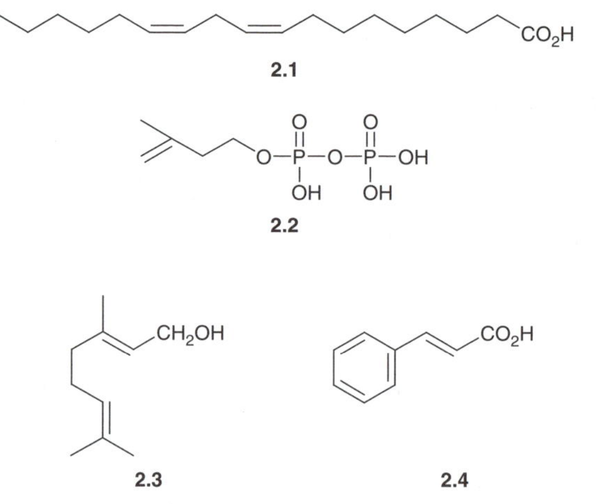

Amino acids are not only the constituents of peptides and proteins and as such are considered as primary metabolites, but some are also precursors of the nitrogenous bases known as alkaloids. Amino acids such as ornithine, lysine, tyrosine and tryptophan play this role. In most alkaloids the carboxyl group of the amino acid has been lost but it is usually possible to discern the remainder of the amino acid. Thus the five-membered pyrrolidine ring of nicotine **2.5**. is derived from ornithine. There are also some specialized amino acids such as alliin (*S*-propenylcysteine) **2.6** which are found in particular plants, in this case onions and garlic.

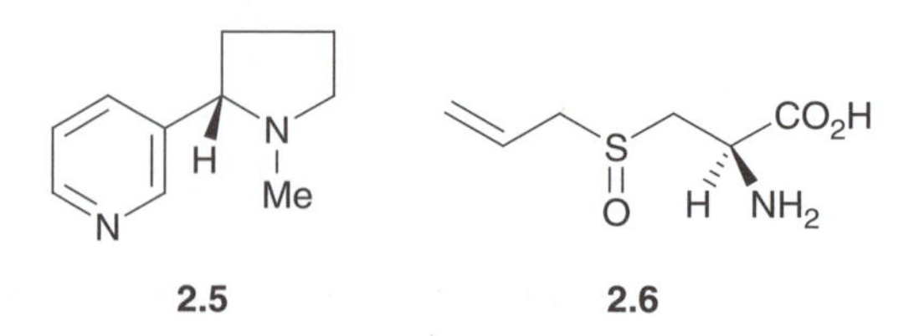

Although simple sugars such as glucose and ribose are primary metabolites, other sugars lacking hydroxyl groups or bearing a methoxyl group are of more limited provenance and can be considered as secondary metabolites. Some of these less common sugars may be attached to a natural product as a glycoside. The non-sugar portion, which may be a polyketide, terpenoid, phenylpropanoid or alkaloid, is known as the aglycone. The presence of the polyhydroxylic polar sugar unit attached to the natural product can confer water solubility. This alters the transport properties of the aglycone and can make a marked difference to its biological properties. The sugar may facilitate the transport of a substance from one part of the plant to another. It may also mask the biological activity which may only become apparent when the glycoside is cleaved. Glucose forms an integral part of the glucosinolates that contribute to the taste of many members of the Brassicaceae which are used as vegetables. The glucosinolate is cleaved by an enzyme myrosinase, which releases an isothiocyanate.

Some natural products are of mixed biosynthetic origin comprising, for example, a polyketide or a terpenoid portion and a phenylpropanoid unit. Large families of alkaloid also contain a terpenoid unit, whilst the anthocyanin plant pigments are derived from the combination of a phenylpropanoid unit and acetate units.

When a compound is biosynthesized by a plant it is often possible to discern a number of stages. The first stage involves the assembly and linking together of the building blocks such as the C_5 isoprene units of the terpenes to form the underlying carbon skeleton. Thus the parents of the terpenes may contain ten, fifteen, twenty or more carbon atoms in a chain. The second stage involves the construction of various ring systems from this chain by cyclizations and sometimes skeletal rearrangements. The third stage involves the stepwise hydroxylations, oxidations and other modifications that lead to the individual natural products. The sugars of the glycosides may be added at this stage. Some hydroxyl groups may be esterified with simple carboxylic acids such as acetic, benzoic, or cinnamic acids. Finally once a natural product has played its specific role in the development of the plant, it undergoes biodegradation to smaller fragments and eventually to carbon dioxide. When the constituents of a plant are examined, compounds may be identified representing these various stages particularly the later stages involved in the various skeletal and oxidative modifications.

The structural relationships between natural products has led to the prediction of various biosynthetic schemes. Over the last fifty years experimental evidence to substantiate this biogenetic speculation has been obtained. A number of plants such as the daffodil, the poppy and various mints, as well as some fungi that are found in the garden, have played an important role in these studies.

2.1 POLYKETIDES

Within each of these groups of natural products it is possible to discern subsets which have structural relationships. The polyketides, particularly those of fungal origin, are grouped in terms of the number of acetate units that form the carbon skeleton. Thus there are tri- (C_6), tetra- (C_8) and penta- (C_{10}) and higher ketides. Characteristic features of the polyketides are the preponderance of compounds with an even number of carbon atoms and oxygenation on alternate carbon atoms. The original idea that polyphenols such as orcinol **2.7** were biosynthesized by the cyclization of polycarbonyl compounds was proposed by Collie in 1907 on the basis of model experiments. The ideas were developed and experimental biosynthetic evidence was provided by Birch in 1953 and subsequently by many others. There is a reductive step in the assembly of the fatty acids leading to saturated chains. The formation of mammalian fatty acids from acetate units was established by Lynen in 1945. In recent years the mechanism of action of the polyketide and fatty acid synthases have been the subject of thorough study. Some unsaturated fatty acids (*e.g.* **2.1**) and polyacetylenes are then derived from these chains by dehydrogenation.

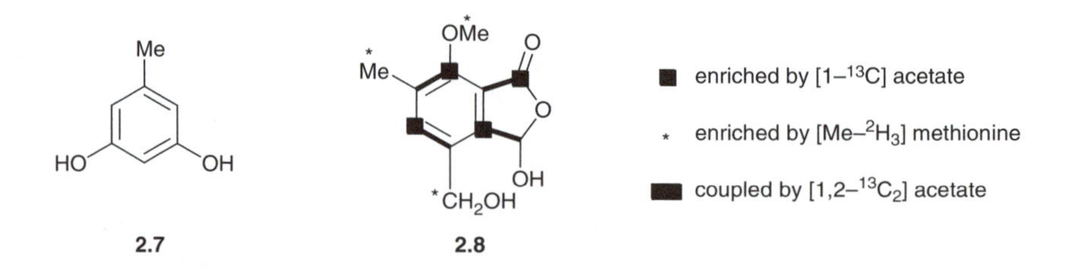

A typical incorporation pattern of acetate units into a fungal metabolite produced by *Penicillium gladioli*, and which was established by carbon-13 labelling studies, is shown in **2.8**. This labelling pattern is consistent with a tetraketide biosynthesis in which the additional carbon atoms were shown to be derived from the methyl group of the amino acid methionine, a common biological methylating agent. The corresponding aldehyde **2.8** is fungistatic and inhibits the growth of other fungi.

2.2 TERPENOIDS

The terpenoids are grouped by the number of C_5 isoprene units that they contain. Thus there are the monoterpenoids (C_{10}), sesquiterpenoids (C_{15}), diterpenoids (C_{20}), sesterterpenoids (C_{25}), triterpenoids (C_{30}) and carotenoids (C_{40}). Natural rubber is a polyterpenoid. The monoterpenoids include many of the fragrant constituents of plants such as geraniol **2.3** from rose oil and menthol **2.9** from mints. The sesquiterpenoids include the bitter principles such as santonin **2.10** from wormwood, whilst the diterpenes such as abietic acid **2.11** are found in wood resins. The triterpenes such as β-amyrin **2.12** are also found in wood resins. The tetracyclic triterpenoids are also the precursors of the steroids. The carotenoids such as β-carotene **2.13** are plant pigments, Many triterpenes and sterols occur in plants with a sugar unit attached to them. In this case they are known as the saponins and the aglycone without the sugar is described as the sapogenin. The combination of a polar hydrophilic sugar with a hydrophobic molecule gives rise to a compound with detergent properties. The word 'saponin' comes from the Latin sapo for soap. The saponins give rise to the foaming of plant wounds surrounding sites of insect attack.

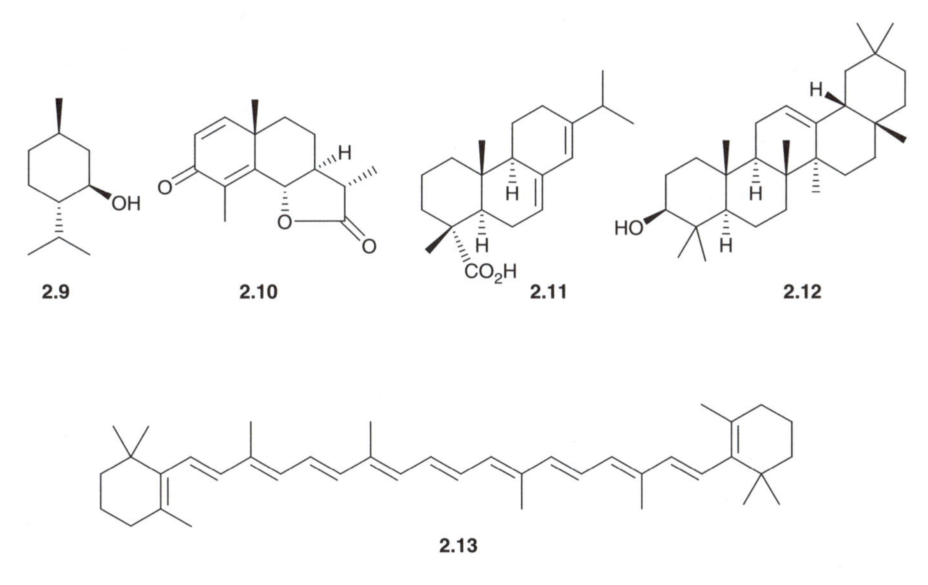

Ideas concerning the origin of the C_5 structural unit of the terpenoids were discussed in the late nineteenth century. The isoprene rule, which stated that the carbon skeleton of the terpenoids were assembled from C_5 isoprene units linked in a head to tail manner, was developed as an aid to structure elucidation by Ruzicka in the 1920s. There are now known to be two major pathways that lead to the C_5 building block, isopentenyl pyrophosphate **2.2**. The first, discovered in 1956, is based on mevalonic acid **2.14** and it was originally explored by Cornforth in the context of the mammalian biosynthesis of cholesterol. Subsequently the existence of this pathway was demonstrated in the biosynthesis of fungal and plant terpenes. The mevalonate pathway proceeds from acetate units via hydroxymethylglutaryl coenzyme A to isopentenyl pyrophosphate. The alternative pathway, sometimes called the deoxyxylulose monophosphate pathway **2.15** was originally discovered in 1996 in a *Streptomycete* species and it has now been found in many higher plants. In plants the mevalonate pathway leading to terpenoids is found in the cytosol, and the non-mevalonate pathway occurs in the plastids. However there is some 'crosstalk' between the pathways. A garden weed, Thale Cress (*Arabidopsis thaliana*), has played a role in these studies. Whereas the formation of carbon–carbon bonds in polyketide biosynthesis is based on carbanion chemistry, the formation of carbon–carbon bonds in the isoprenoid chain utilizes carbocations based on the allylic pyrophosphate acting as a leaving group.

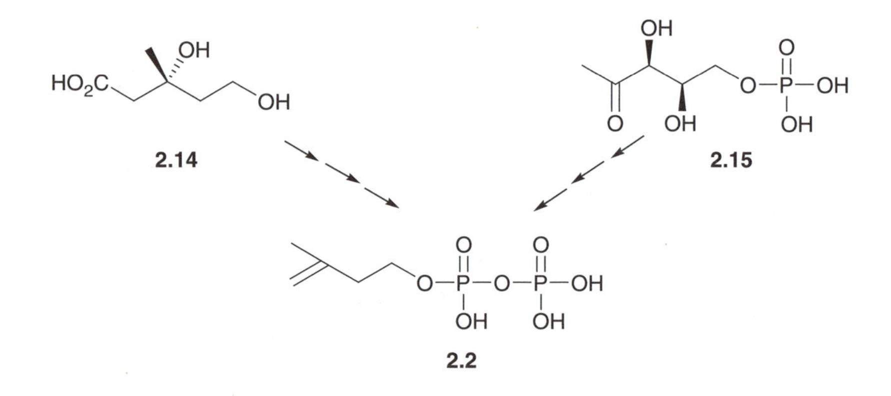

The enzymatic chemistry of terpenoid cyclizations has been thoroughly studied. Many of the cyclizations that lead to the mono- and sesqui-terpenoids are initiated by the pyrophosphate group generating a carbocation which then reacts with the alkenes in the carbon chain. On the other hand the biosynthesis of the higher terpenoids involve acid-catalysed polyene cyclizations. In mammals and fungi the formation of tetracyclic triterpenes and thence the steroids proceeds from squalene epoxide **2.16** via lanosterol **2.17**, whilst in plants a different compound, cycloartenol **2.18**, which contains a cyclopropane ring, is the first triterpene to be formed. The presence of the products of particular terpenoid cyclizations can have taxonomic significance and this will be mentioned later. Lanosterol is the precursor of many sterols. Its conversion into the fungal sterol ergosterol is the target of the azole and morpholine fungicides such as triadimefon® and tridemorph®.

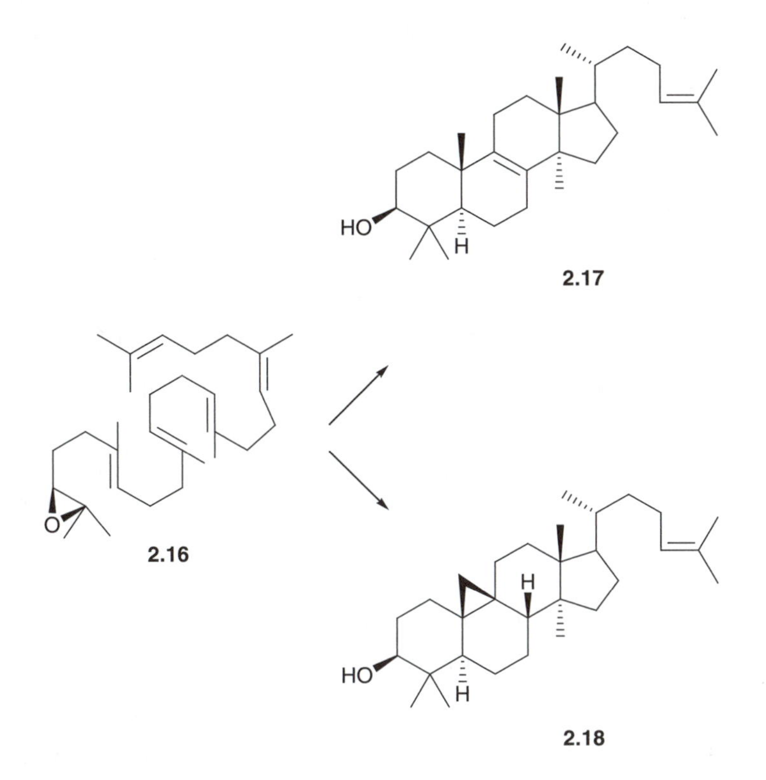

2.3 PHENYLPROPANOIDS

The phenylpropanoid family of natural products are those that contain a C_6–C_3 unit such as cinnamic acid **2.4** and various coumarins. The biosynthetic pathway leading to the C_6–C_3 pathway is known as the shikimic acid pathway. This pathway falls into two sections. The first section involves the formation of shikimic acid **2.22** from phosphoenol pyruvate **2.19** and erythrose 4-phosphate **2.20** via the phosphate of a C_7 acid, heptulosonic acid **2.21**. The second stage involves the addition of a further phosphoenol pyruvate **2.19** to shikimic acid **2.22** to give chorismic acid **2.25**. This undergoes a Claisen rearrangement to prephenic acid **2.24**. Decarboxylation then forms the aromatic ring of phenylpyruvic acid **2.23**.

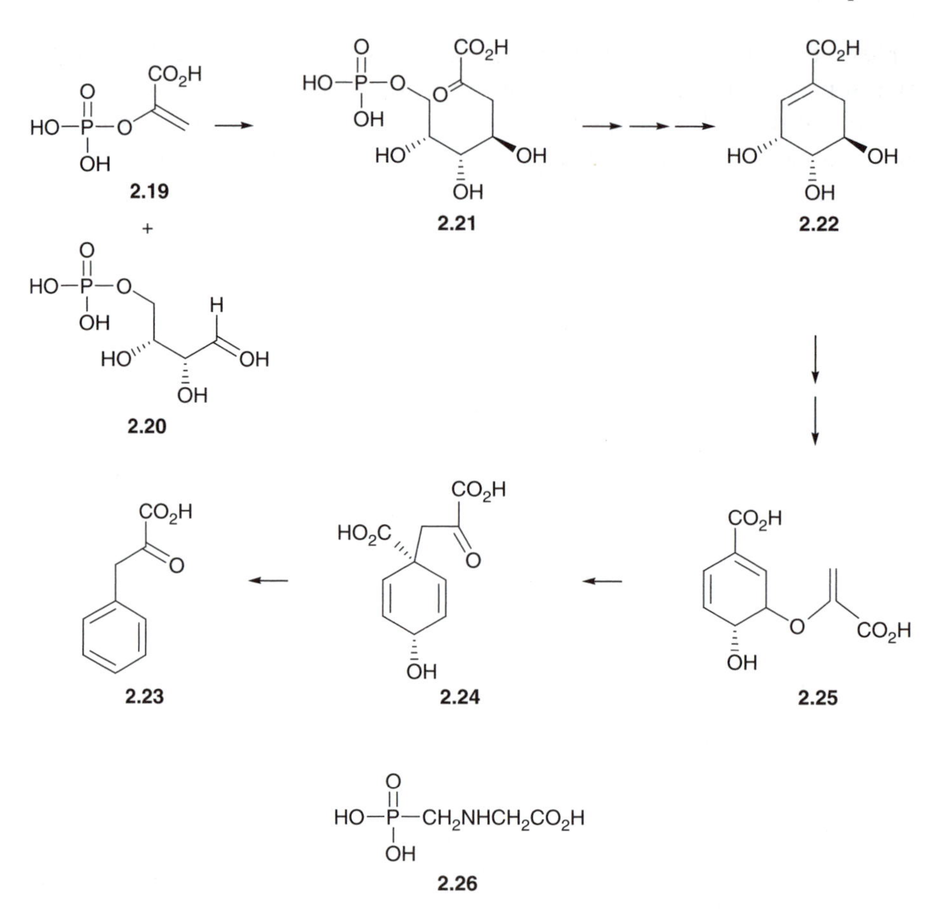

This pathway is not found in man and hence it has made a good target for herbicides. The C_6–C_3 amino acids such as phenylalanine which we require, are obtained from the diet. A key enzyme in the C_6–C_3 pathway in plants is the 5-enolpyruvyl-shikimate-3-phosphate synthase (EPSP) which catalyses a step between shikimic acid **2.22** and chorismic acid **2.25**. This enzyme is the target for the herbicide glyphosate (Roundup®) **2.26**.

A further series of compounds are derived from the combination of a phenylpropanoid unit and a polyketide, typically a triketide. This gives rise to the chalcones (*e.g.* butein **2.27**), flavanols (*e.g.* quercetin **2.28**) and anthocyanidins (*e.g.* delphinidin **2.29**). Butein is found in the bark of various trees such as *Acacia* species, whilst glycosides of the flavonols and anthocyanidins form the pigments of many flowers. The aromatic amino acids, phenylalanine **2.31**, tyrosine **2.32** and indirectly tryptophan **2.33**, which lead to the alkaloids, are C_6–C_3 phenylpropanoids.

Although some aromatic natural products are formed by the dehydrogenation of alicyclic rings, many arise via the polyketide or phenylpropanoid pathways. Aromatic natural products formed by the polyketide pathway often have oxygen atoms on alternate carbon atoms.

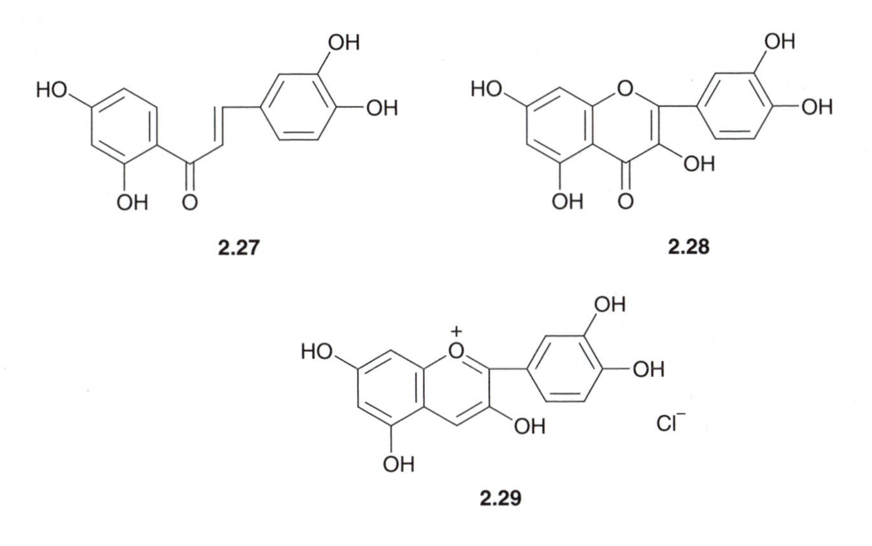

2.4 ALKALOIDS

Although the alkaloids may be grouped in terms of their building blocks, such as those that are derived from ornithine **2.30** and tyrosine **2.32** and those that are derived from tryptophan **2.33**, the classification is more commonly in terms of the family of plants from which the alkaloids are isolated (*e.g.* the alkaloids of the Amaryllidaceae) or on the basis of some structural characteristic such as the benzylisoquinoline or indole alkaloids. This classification pre-dates the accumulation of the biosynthetic evidence for the origins of these natural products.

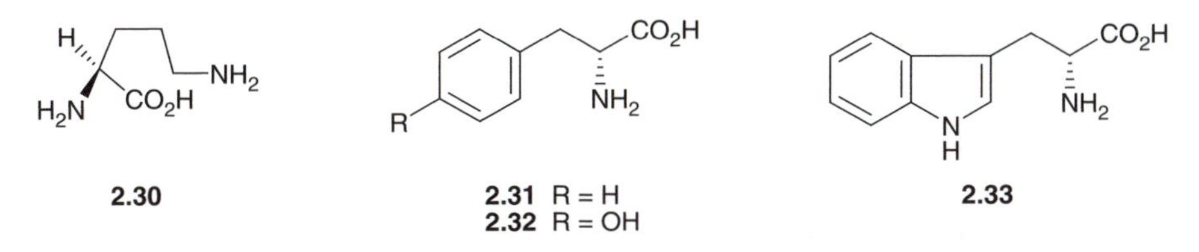

Nicotine **2.5** is found in various *Nicotiana* species. Whilst the pyrrolidine ring is derived from the amino acid ornithine **2.30**, the pyridine ring has a slightly more complicated biosynthesis from aspartic acid and glyceraldehyde.

A number of alkaloids which are biosynthesized from the aromatic amino acid tyrosine **2.32** occur in garden plants including daffodils, autumn crocus, poppies and *Berberis* species. Their biosyntheses illustrate several key biosynthetic reactions particularly the role of radical reactions in phenol coupling. An aromatic amino acid such as tyrosine **2.32** undergoes decarboxylation to a C_6–C_2 phenylethylamine. This can also form an aldehyde by transamination. Condensation between the aldehyde and the amine and further modification affords benzylisoquinolines such as the alkaloid papaverine **2.34** which is found in poppies (*e.g. Papaver orientale* and *P. somniferum*). A similar condensation and reduction can take place between a benzaldehyde and tyramine. Particular hydroxyl groups are blocked as their methyl ethers, whilst others are left free, as in reticuline **2.35**, to participate in the enzymatic phenol coupling reactions. A key phenol coupling step which occurs in the oriental poppy, *P. orientale*, and the opium poppy, *P. somniferum*, involves the conversion of (*R*)-reticuline **2.36** to salutaridinone **2.37**. Morphine **2.38** is formed as a result of further modifications in the opium poppy. Ortho:para phenol couplings of norbelladine derivatives **2.39** lead to the alkaloids galanthamine **2.40** and lycorine **2.41** which are found in daffodils. Berberine **2.42**, which is found in *Berberis*, contains an extra carbon atom (the 'berberine bridge'), which is derived by the oxidation of a methyl group.

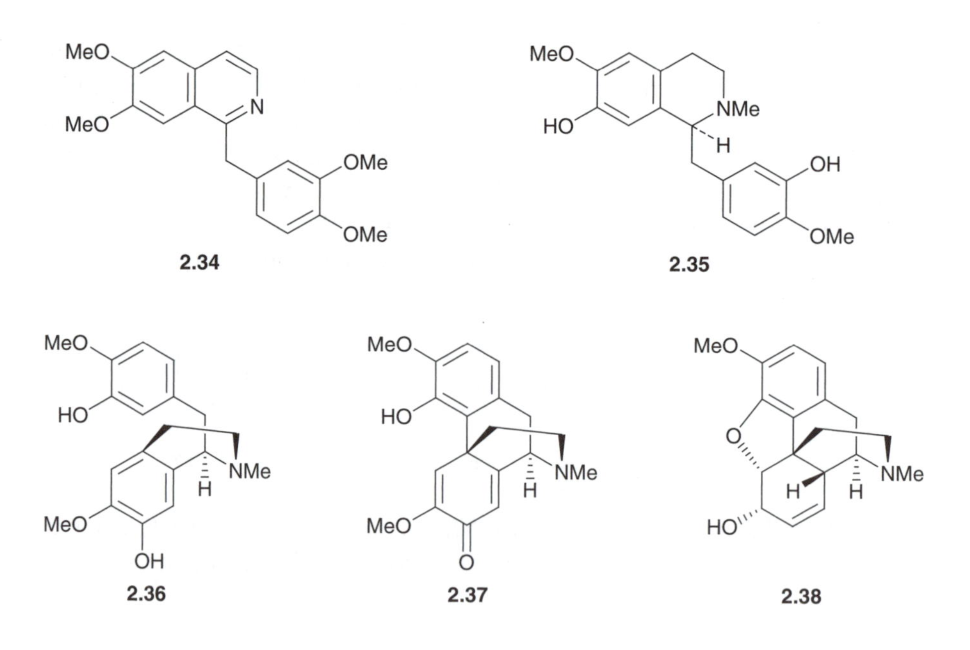

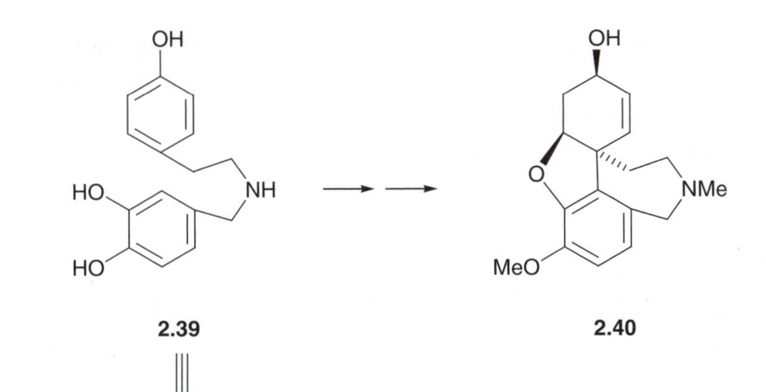

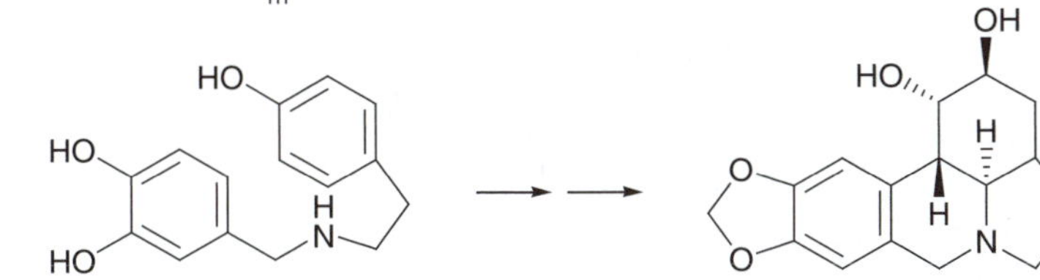

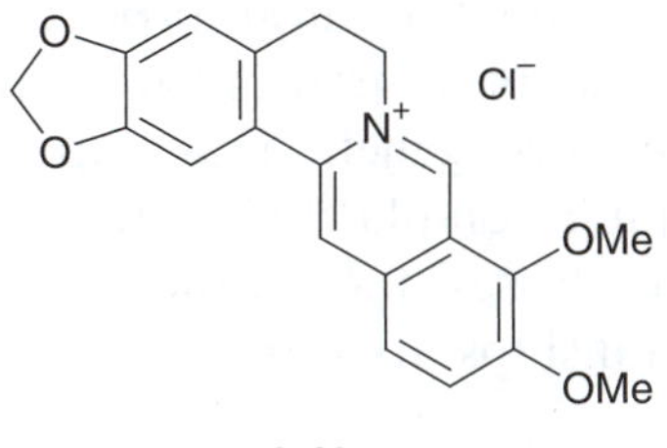

2.42

The way in which these experiments on the biosynthesis of plant constituents were carried out in the 1960s was to synthesize a radioactive sample of the precursor bearing a carbon-14 or tritium label at a specific site in the molecule and then feed this to the plant. The feeding experiments involved attaching a small sample tube, containing a solution of the precursor, to the stem of the plant. A cotton wick was threaded through the stem and dipped into the solution of the precursor. After the solution had been taken up by the plant, the plant was harvested. The metabolites were isolated and the site of the radiochemical label was established by chemical degradation. More modern methods use enzyme preparations and substrates labelled with the stable isotopes, carbon-13 and deuterium. The sites of labelling can then be established by NMR and mass spectrometric methods.

Many alkaloids contain structural fragments that resemble mammalian neurotransmitters such as acetylcholine **2.43**, dopamine **2.44**, adrenalin **2.45** and serotonin **2.46**. Not surprisingly some of these alkaloids bind to the cholinergic, dopaminergic, adrenergic or 5-HT (5-hydroxytryptamine) receptors and can have significant biological effects as psychotropic or neurotoxic agents.

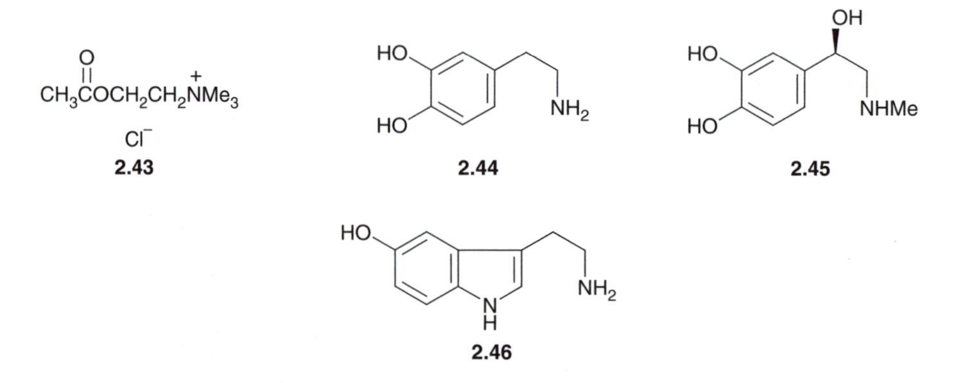

A feature, which may already be apparent from this discussion of natural product biosynthesis, is that related species of plants often produce structurally related natural products. Chemosystematics is one of the tools that has been used in the classification of plants since it in turn must reflect their genetic make-up. In the later chapters we shall come across quite complicated structures. By looking at these through 'biosynthetic glasses' their structures can be rationalized and their biogenetic relationships discerned.

CHAPTER 3

Natural Products and Plant Biochemistry in the Garden

In this chapter we shall consider some of the general structural materials and biochemical processes that occur in garden plants. A number of garden agrochemicals interact with these processes and the biochemicals that are involved.

3.1 THE STRUCTURAL MATERIAL OF PLANTS

Plant cell walls have a number of functions in which the chemistry of their components plays an important role. The cell walls provide a mechanical strength and some plasticity to prevent the cell from rupturing as a consequence of the internal cellular chemistry. They provide both the containment for the transport of water and nutrients and the prevention of dehydration, and yet they must retain some permeability. They need to provide protection against insects and plant pathogens and contain substances that absorb ultraviolet light and prevent photochemical damage to the cellular contents.

Cellulose is one of the most abundant bio-organic polymers within the cell wall. It is a β-1,4′-polyacetal of the disaccharide cellobiose **3.1**. The number of units in the polymer may be as high as 10 000–15 000. Because of the many opportunities for both intramolecular and intermolecular hydrogen bonding between the sugar units, the network that this creates hinders free rotation and imposes a rigidity on the structure of cellulose. Since the bonding between the sugars involves equatorial rather than axial groups, the polysaccharide adopts an extended linear conformation to give fibrous structures. This provides the basis for the function of cellulose as a component of the structure of the leaves and other parts of the plant.

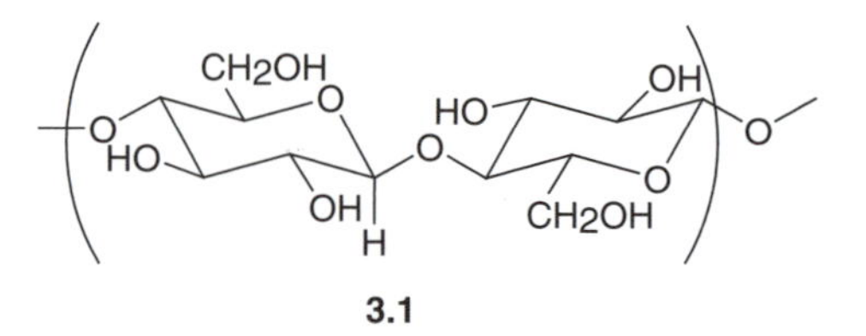

3.1

The primary cell wall contains cellulose microfibrils and some glycoproteins embedded within this matrix. These structural glycoproteins have characteristic amino acid sequences. They include the extensins with regular tyrosine units, and hydroxyproline-rich and glycine-rich proteins. The carbohydrate side chains attached to the hydroxyproline comprise galactose and 1,4-arabinose residues. The presence of these amino acid sequences confers specific properties on the cell wall in terms of resistance to attack. For example there is evidence that cross-linking between extensins involving the tyrosine units takes place when a cell wall is damaged. Root nodule extensins in leguminous plants play a role in the attachment and development of the symbiotic *Rhizobium* species of nitrogen fixing bacteria.

Although the polymeric structure of cellulose precludes water solubility, the extensive hydrogen bonding network and the presence of additional sites for hydrogen bonding mean that cellulose can absorb a significant amount of water. This interaction with water confers both advantages and disadvantages to the plant. Many leaves are covered with a hydrocarbon wax to reduce both waterlogging of the leaf when it rains and excessive water loss under arid conditions. This wax also behaves as a lubricant to make freshly fallen wet autumn leaves slippery. Biodegradation of the wax exposes the cellulose on the leaf which can then become waterlogged.

Hydrophobic wax coatings are found on other parts of garden plants, for example on the tuberous roots of vegetables such as the potato. The wax coating of fruit is an important protection against dehydration and microbial spoilage. Apples with a good wax coating keep longer. There was an old method of storing apples which used an oiled paper wrapping. Today some fruit are sprayed with wax to preserve them.

Whereas cellulose is a structural polysaccharide, starch is a storage polysaccharide. Starch is the major storage polysaccharide of many seeds and tubers, for example, the potato. It differs from cellulose by containing α-1,4′-linked D-glucose units **3.2**. These form a polymer known as α-amylose. The other major component of starch is amylopectin, which contains the α-1,4′-linked D-glucose backbone together with some chain branching. In contrast to cellulose which has a fibrous structure, the α-amylose of starch with an α-1,4′ (axial:equatorial) linkage is wound into a more compact structure. Different enzyme systems are involved in the cleavage of cellulose and starch leading to differences in the ease with which they are digested.

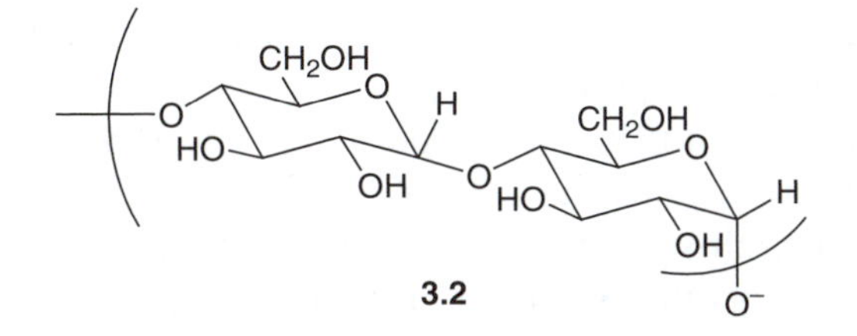

3.2

Inulin is another storage polysaccharide for some plants, particularly members of the Asteraceae (Compositae) such as chicory (*Cichorium intybus*), artichokes (*Helianthus tuberosus*) and dandelion (*Taraxacum* spp.). Unlike starch, inulin is a fructan composed of fructose units. The presence of the different fructose β[2–1] linkage means that it is not easily hydrolysed in the stomach but it is broken down by bacteria in the intestine. The bacterial fermentation by *Bifidobacteria* sp. and other organisms, favours those organisms which produce short chain fatty acid such as acetic, propionic and butyric acids together with methane and hydrogen. The formation of the gases accounts for some of the discomfort and flatulence associated with these foodstuffs. However there are beneficial effects from the presence of inulin and the fact that it is not hydrolysed to sugars. For example it does not raise blood sugar levels and hence it can be of value in diabetic foods.

When a tree or a fruit is damaged, it may produce a gummy exudate to protect the site of injury. These plant gums are polysaccharides which possess a branched chain structure often containing a glucuronic acid **3.3**. A typical plant gum might contain a core of β-1,3-D-galactose units **3.4** to which side chains of different sugars such as L-arabinofuranose, L-rhamnopyranose and D-glucuronic acid are attached. This branched structure reduces the tendency of the gum to crystallize and, because there are different sugars requiring different enzymes to metabolize them, it also reduces the sensitivity of the gum to microbial enzymatic attack. The mucilage around seeds also contains polysaccharides. The hydrogen bonding properties of the sugar units help to retain water and protect the seeds from dessication.

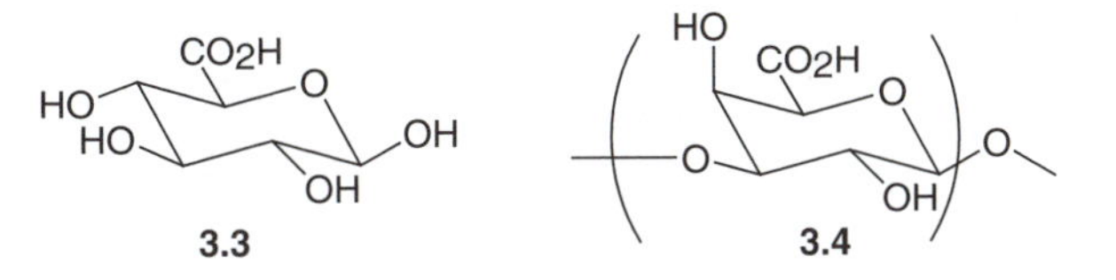

3.3 **3.4**

The cell wall also contains pectic acids and pectins. The pectic acids are polygalactouronic acids in which the chains are bridged by magnesium and calcium ions. The pectins, which are also formed in the primary cell wall of some fruits such as apples, contain D-galactouronic acid units that are partially methylated. These provide the basis of the extensive gelling properties of the pectins which occur in the presence of plant acids and sugar in jam making.

As a plant ages, lignin begins to permeate the polysaccharide membrane. Lignin comprises aromatic C_6–C_3 units and it is structurally quite different from cellulose. Lignin is formed with building blocks of 4-hydroxyphenylpropenol **3.5**, coniferyl alcohol **3.6** and sinapyl alcohol **3.7**. The polymerization process is a phenol coupling reaction based on the radicals shown in **3.8**. It is mediated by iron-containing haem enzymes and leads to structures which include units such as **3.9**.

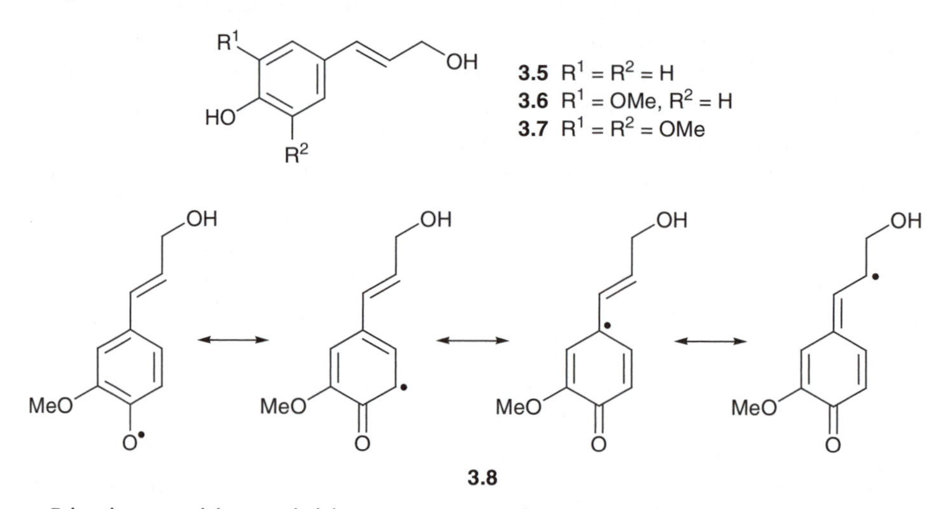

Lignin provides a rigid structure to plant material. Softwood lignins are mainly assembled from coniferyl alcohol. On the other hand hardwoods are assembled from coniferyl and sinapyl alcohols. The lignins in grasses and their relatives, such as the tall ornamental grasses including bamboos, contain all three structural units.

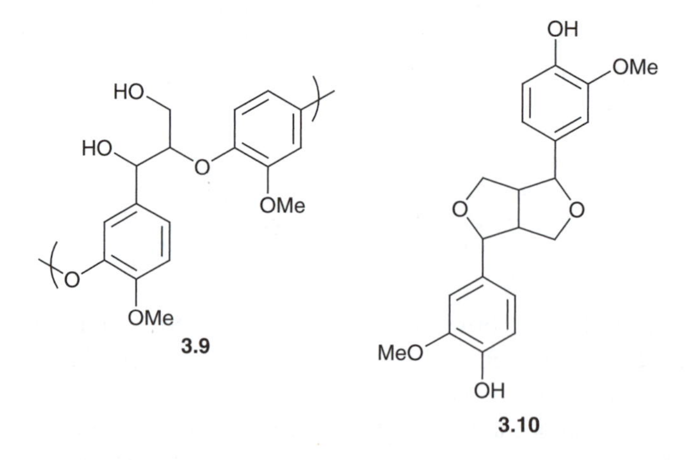

The free radicals are derived from compounds such as coniferyl alcohol **3.8** and have several potential sites for phenol coupling reactions allowing for cross coupling to occur, imparting strength to the wood. These enzymatic phenol coupling reactions also lead to low molecular weight compounds known as lignans and are exemplified by pinoresinol **3.10**. Some of these, such as podophyllotoxin and the analogue etoposide, have attracted interest because of their cytotoxic activity as potential anti-tumour agents.

Many woods, such as oak, also contain phenols which have a limited solubility in water. These phenols bind to the proteins in, for example, leather and are used to tan leather, hence the name tannins for these compounds. These phenols also have anti-bacterial properties accounting for some of the use of oak casks in storing wine and beer. Some of the simpler gallotannins are glucose esters of gallic acid **3.11** or polygalloyl units in which the galloyl units are linked as phenol esters. The more complex ellagitannins are glucosides of ellagic acid **3.12**. Ellagic acid is a dimer of gallic acid which is formed by a phenol coupling reaction. The preservative action of these phenols in wood is supplemented when the fence is creosoted. Creosote is a mixture of phenols including the cresols.

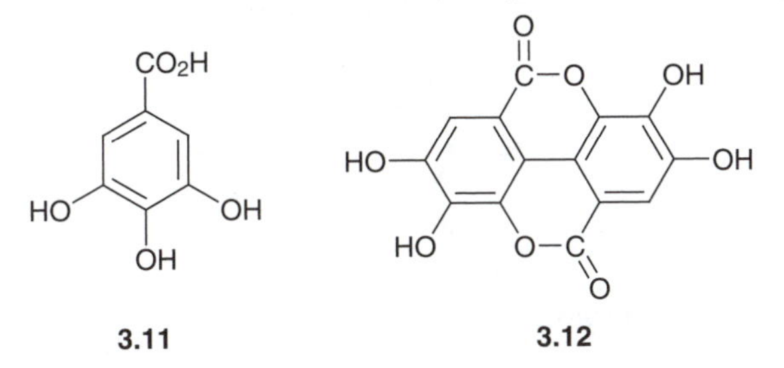

Cellulose is degraded by cellulytic enzymes that occur in various bacteria, fungi, insects and molluscs such as snails. The lignases, the enzymes required to degrade lignin, are less widespread in micro-organisms than cellulolytic enzymes. They are present in the 'white-rot' wood-rotting fungi such as *Ganoderma, Polyporus* and *Trametes* species. This difference in biodegradative attack accounts for the skeletal appearance of some fallen leaves in which the cellulose has been degraded before the lignin. Holly leaves are particular prone to this. The white-rot organisms may be found not only on trees in the garden but also on garden fences and posts. The white-rot mycelium can often be seen as the rotten wood is prized apart. Degradation by lignases not only allows the mycelium and moisture to penetrate the wood but it also allows insects such as woodlice to enter. Some lignin-degrading fungi such as *Phanerochaete chrysosporium* and *Trametes hirsuta* will degrade the wood preservatives, creosote and pentachlorophenol.

3.2 PHOTOSYNTHESIS

Sunlight is trapped to provide the energy for plant growth and development by a process known as photosynthesis. This takes place within the chloroplasts of the leaf. These organelles contain the green pigments, chlorophyll a **3.13** and chlorophyll b and some carotenoids such as β-carotene and lutein. Chlorophyll b contains an aldehyde in place of a methyl group in the chlorin (dihydroporphyrin) ring and thus it has a slightly different absorption spectrum to chlorophyll a. However both absorb light in the blue and red regions of the spectrum and hence reflect the green light. The name chlorophyll was given to the green pigment of plants by Pelletier and Caventou in 1817 and chlorophyll a and b were purified by Willstatter in 1906–1914. Subsequently they were separated chromatographically in 1933. Their structure was established as a result of the work of Willstatter and then Fischer in the 1930s.

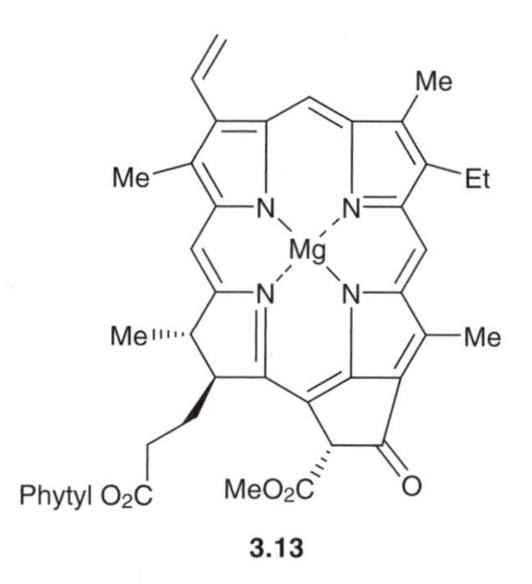

3.13

Chlorophyll is degraded quite rapidly. Throughout the spring and summer there is a continuous process of degradation and resynthesis in the leaf. The central magnesium is quite labile. Its removal gives the pheophytins a and b a brownish colour. When the lawn goes brown in a dry summer, the synthesis of chlorophyll has ceased. There is no water to move the products of photosynthesis around the plant. Provided the relevant enzyme systems have survived, the biosynthesis of chlorophyll resumes when it rains, and moisture brings nutrients to the leaf and normal metabolism can take place again. When green vegetables are boiled, the magnesium may be eluted from the chlorophyll particularly if the water is slightly acidic. Cooked cabbage can loose its colour becoming quite pale or even slightly brown in colour. The addition of small amounts of sodium bicarbonate to neutralize the plant acids can reduce the colour loss. This loss of colour is a problem in canning peas. The green colour was sometimes supplemented by the addition of sodium copper chlorophyllin in which the copper complex was more stable.

The carotenoids and the anthocyanins play a protective role in the leaf, acting both to absorb some ultraviolet radiation and as anti-oxidants.

Photosynthesis has two stages. The first stage, the light reaction, is initiated by the absorption of light by the chlorophyll and the consequent promotion of an electron to an orbital of higher energy. The effect of this promotion is transferred through protein:pigment complexes known as photosystems I and II. This energy is used to split water and release oxygen at an oxygen evolving centre (the Hill reaction). The absorption of light also leads to the generation of ATP and the provision of electrons to reduce NADP to NADPH. In the second, dark reaction these two reagents are used with the intervention of the sugar, ribulose 1,5-diphosphate **3.14**, to fix carbon dioxide leading to the formation of glyceraldehyde-3-phosphate **3.15** and then the sugar, fructose-6-phosphate **3.16** from which glucose is derived.

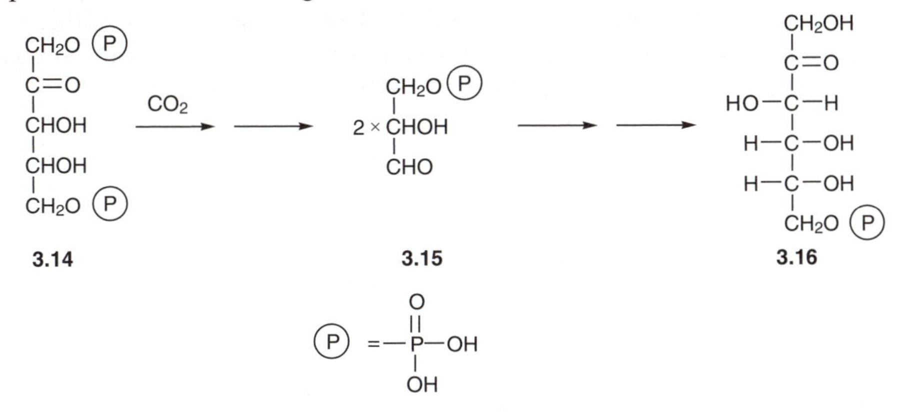

The fixation of carbon dioxide is a cyclical process (the Calvin cycle) and thus it requires the regeneration of the C_5 sugar, ribulose diphosphate. In this cycle the formation of six molecules of fructose 6-phosphate requires six molecules of the C_5 sugar, ribulose diphosphate, and six molecules of carbon dioxide. Of the six molecules of fructose 6-phosphate, five of these C_6 molecules are recycled into six C_5 molecules of ribulose diphosphate. The remaining one molecule of fructose-6-phosphate is converted into glucose for use in general metabolism as an energy store and as a biosynthetic starting material. The degradation of glucose provides primary building blocks such as acetyl coenzyme A for polyketide, fatty acid and isoprenoid biosynthesis.

Plants which fix carbon dioxide by this cycle are know as C-3 plants because the primary products of photosynthesis are C_3 molecules. Some plants such as tropical grasses which grow in regions of high sunlight, have an alternative pathway in which a C_3 sugar acts as the receptor for carbon dioxide and a C_4 sugar is the primary product of photosynthesis. These plants are known as C-4 plants. An interesting difference between the two pathways is that, because of isotope effects, the ^{12}C:^{13}C ratio in the products, differs slightly. Sucrose from sugar cane (a C-4 plant) has a slightly different ^{12}C:^{13}C ratio to the sucrose derived from sugar beet, a C-3 plant. Thus the origins of sugars and their metabolites can be detected by careful isotope ratio measurements.

A number of herbicides target the redox processes involved in photosynthesis. These include the bipyridinium herbicides paraquat **3.17** and diquat and the triazines such as atrazine **3.18**, and the nitrodiphenyl ethers. The dipyridyl dication of paraquat is easily reduced to a resonance stabilized radical cation in the presence of light in the chloroplast. This converts oxygen to a reactive and destructive superoxide anion. As an ionic species, paraquat is deactivated on contact with the soil. A number of other herbicides, such as the arylureas, the dinitrodiphenyl ethers and the triazines such as atrazine **3.18**, are sufficiently stable in soil to be taken up by emergent weed seedlings. These compounds inhibit the Hill reaction and photosynthetic electron transport.

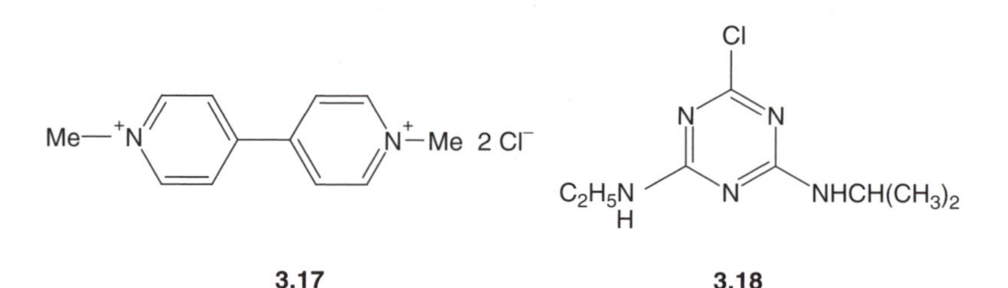

Light is required not only for photosynthesis but also for other aspects of plant growth and development including photoperiodism. Some of these responses are to red light whilst others are blue light dependent. The red light responses are mediated by a protein: linear tetrapyrrole complex known as phytochrome. The phytochrome photoreceptor has two interconvertible forms, one absorbing around 666 nm (red) and the other around 730 nm (far-red). The interconversion involves a double bond isomerisation. Some biological processes such as seed germination are dependent on the presence of red light. Certain seeds that are buried too deep in the soil do not receive enough light to germinate. Other processes such as stem elongation are modified by the ratio of the red to far-red forms of phytochrome. The formation of rather straggly plants in shady positions is related to this. The induction of flowering and the long-day or short-day responses are also related to the interaction of light with the phytochrome system.

3.3 OXIDATIVE COENZYMES

The porphyrin ring system also forms part of the haem complexes that contain iron as the central metal. These coenzymes are involved in various oxidative and electron-transport processes in which the variable valency of iron plays a central role. The enzymes include the cytochrome P_{450} mono-oxygenases which insert oxygen into a molecule, and the haem peroxidases which catalyse the hydrogen peroxide dependent oxidation of a wide variety of substrates. The haem peroxidases include horseradish peroxidase, lignin peroxidase and ascorbate peroxidase. The free radicals that are generated by these systems from their various, often phenolic, substrates undergo dimerization and polymerization reactions.

The later stages of many biosyntheses involve a hydroxylation. These are often catalysed by a cytochrome P_{450}. An example, found in a number of biosyntheses, involves the hydroxylation of a methyl group in the sequence $CH_3 \rightarrow CH_2OH \rightarrow CHO$. The removal of a steroidal C-14α methyl group **3.19** during the biosynthesis of the fungal cell wall component ergosterol and the oxidation of the C-19 methyl group of ent-kaurene **3.20** in gibberellin plant hormone biosynthesis are two examples. The nitrogen in a five-membered heterocyclic azole ring co-ordinates to iron and will therefore inhibit the oxidation. Azole fungicides such as triadimefon **3.21** act at this stage of ergosterol biosynthesis, whilst another azole, paclobutrazole **3.22**, acts as a plant growth regulator by blocking gibberellin plant hormone biosynthesis.

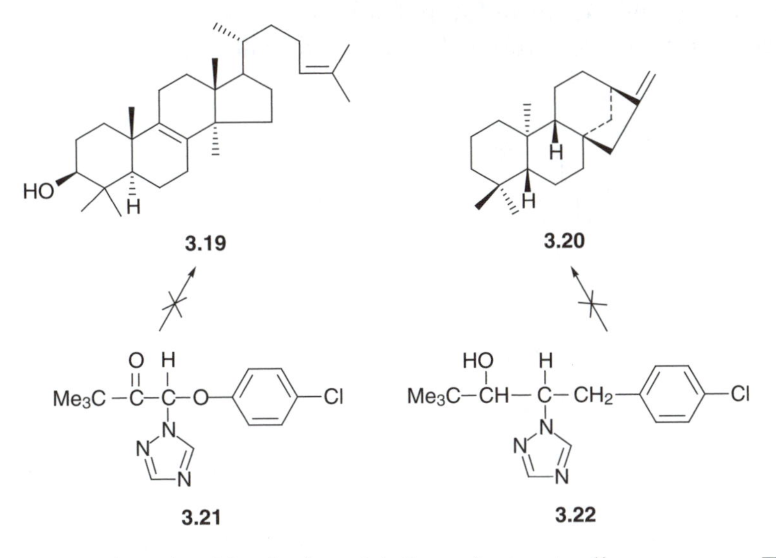

Iron is also involved in the iron(II)-2-oxobutyrate dioxygenases. These enzymes mediate a variety of biosynthetic oxidation reactions in plants including some stages in the biosynthesis of the gibberellin plant hormones, flavonoid biosynthesis and the biosynthesis of ethylene. In the formation and function of these coenzymes and the proteins with which they are associated, we see the need for metal ions as well as phosphorus in phosphates and nitrogen in amino acids. The gardener may have to supply these in fertilizers. An iron deficiency can have quite a serious effect on the development of a plant.

3.4 PLANT HORMONES

The enzyme systems responsible for plant growth and development are under hormonal control. The plant growth hormones are a structurally diverse group of substances that play a role in mediating various aspects of plant growth and development. They are produced by plants in very small amounts, often at the level of μg/kg. Consequently their detection and much of our knowledge of their biological function has rested on instrumental methods of analysis, particularly gas chromatography and mass spectrometry. Some of these compounds or their analogues have been used in agriculture or horticulture to accelerate fruit ripening and to modify plant growth. The biosynthesis of these compounds and the genetics of plant hormone production have been the subject of intense study for over seventy years.

The first plant hormone to be detected was the auxin, indolyl-3-acetic acid **3.23**. It is biosynthesized in the growing tips of plants and it induces pronounced plant growth and stimulates root growth and flowering. It has been claimed that a corn root will respond to as little as 10^{-12} g of indolyl-3-acetic acid.

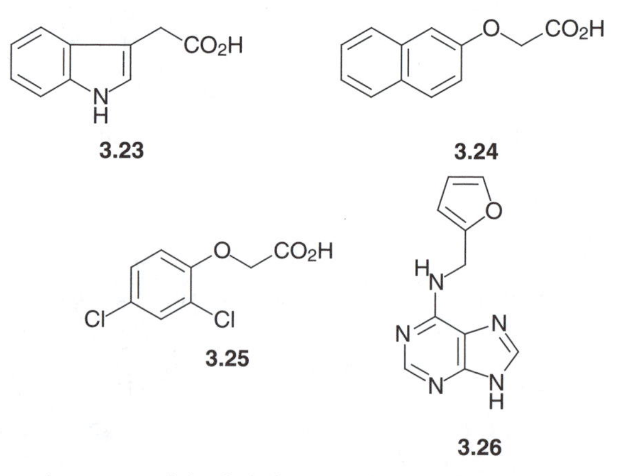

Synthetic analogues of indolyl-3-acetic acid such as 2-naphthoxyacetic acid **3.24**, are used as 'rooting hormone'. It has also been used in a fruit spray to enhance the setting of tomatoes, strawberries and grapes. 2,4-Dichlorophenoxyacetic acid (2,4-D) **3.25** and its relatives such as 2-methyl-4-chlorophenoxyacetic acid (MCPA) are present in 'hormone' weedkillers. Their greater uptake by broad-leafed weeds leads to the rapid growth of these weeds, a disturbance of their growth pattern, and their eventual death. γ-(2,4-Dichlorophenoxy)butyric acid and its 2-methyl analogue have the interesting property of acting as selective herbicides against only those weeds that can degrade them by β-oxidation to the active phenoxyacetic acid. The herbicide α-(2, 4-dichlorophenoxy)propionic acid (mecoprop®) is chiral and one enantiomer is more active than the other. 2,3,6-Trichlorobenzoic acid and 2-methoxy-3,6-dichlorobenzoic acid (dicamba®) also appear to act as indolylacetic acid mimics.

The cytokinins, kinetin **3.26** and the related zeatin and N^6-isopentenyladenine, are formed predominantly in young roots and stimulate cell division. In contrast to the auxins which promote apical dominance, the cytokinins promote the development of side shoots. An alteration in the balance of these may account for the differences between the tall and shrubby varieties of plants such as the tomato. They also have an effect on the development of fruit and there is an application in conjunction with gibberellins A_4/A_7 to improve the development of some apple varieties. Cytokinin analogues have been reported as extending the shelf life of cut flowers.

The gibberellins were originally isolated as the phytotoxic metabolites of a rice pathogen *Gibberella fujikuroi* (*Fusarium moniliforme*). Crude material was isolated in 1938. Much of the chemistry of the gibberellins was established with the fungal metabolite gibberellic acid **3.27** which was isolated in 1954. In the late 1950s it was realised that the phytotoxic effect of the gibberellins involving excessive stem elongation was actually an over-response to a normal hormonal effect. Gibberellins were then found in minute amounts (mg/kg) in peas, beans and, subsequently, in many other plant species. Although they are present in minute amounts throughout the plant, the seeds and fruits have proved to be the best sources of the material.

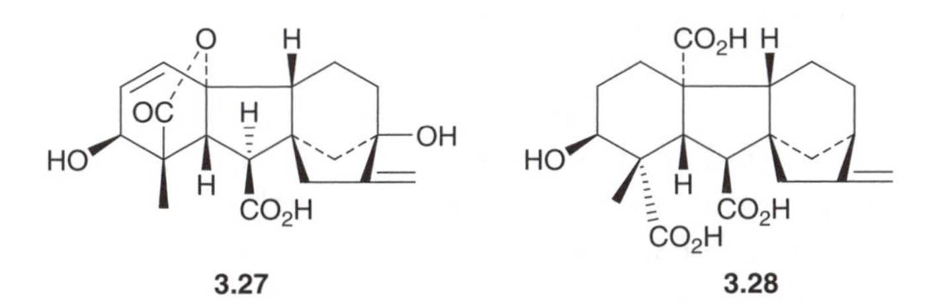

There are about 140 gibberellins that are known, each being designated as gibberellin A_n. They have the same underlying unique carbon skeleton and differ from each other in their hydroxylation pattern. The gibberellins are all carboxylic acids. There are two series of gibberellins. One series *e.g.* GA_{13} **3.28**, retains the twenty carbon atoms of their diterpenoid precursor, whilst the other series has lost one carbon atom and possesses a γ-lactone ring in its place as in **3.27**. The C_{19} compounds are formed from the C_{20} compounds. Biological activity resides mainly with the C_{19} compounds. The best known gibberellin is gibberellic acid (GA_3) which is available commercially and is produced by fermentation.

Examination of any one plant by gas chromatography linked to mass spectrometry may reveal the presence of many gibberellins related to each other in a common biosynthetic sequence. For example twenty-four gibberellins have been identified in apples. The gibberellins occur not just as the free hormones but also as glucosides. Recently evidence has been presented for the existence of a gibberellin receptor.

Gibberellins influence many aspects of plant growth and development. In the seed, gibberellins stimulate the action of α-amylase, releasing sugars from starch. Commercially gibberellic acid is used for this purpose at a concentration of about 0.5 mg/kg of barley in hastening the malting of barley grain in beer manufacture. The gibberellins stimulate longitudinal cell growth and stem elongation. Gibberellins also have an effect on the development of fruit such as apples and citrus fruits. In the fruit the gibberellins inhibit senescence and maintain a healthy skin, reducing the ease of attack of pests. Another commercial application is in a spray for improving the size of seedless grapes. Gibberellic acid can also be used to make ornamental plants flower out of season. Some rosette plants like the lettuce bolt and flower only under 'long-day' conditions, whilst others normally grown as annuals like carrots, will flower as biennials after a cold period. Treatment with gibberellic acid breaks this requirement. Some synthetic plant growth regulators such as paclobutrazole (the plant growth regulator Bonzi®) **3.22** and chlorocholine chloride (CCC) are gibberellin biosynthesis inhibitors reducing stem length. Gibberellin biosynthesis is blocked in some genetically dwarf plants.

A group of plant hormones which have been discovered more recently are the brassinosteroids, exemplified by brassinolide **3.29**. These steroid hormones, of which there are some sixty known compounds, were originally discovered in 1979 in the pollen of oil seed rape *Brassica napus*. They are known to be widespread throughout the plant kingdom. For example they have been found in the pollen of the sunflower (*Helianthus annuus*) and in peas (*Pisum sativum*), beans (*Phaseolus* *vulgaris*) and radish (*Raphanus sativus*). Because they occur in minute amounts, much of their biological activity has been established with synthetic material prepared from readily available plant sterols. The brassinosteroids appear to stimulate a number of aspects of root and stem growth, acting synergistically with indolylacetic acid. They particularly influence a plant under conditions of stress, conferring heat and cold tolerance and resistance to drought. This anti-stress effect may have potential applications in crop production.

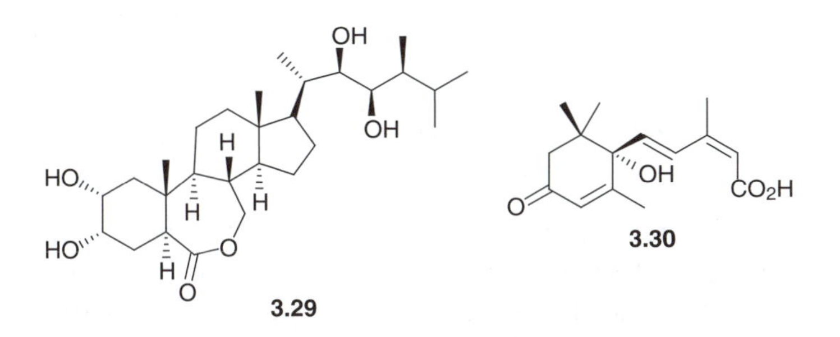

Whereas the plant hormones that have been described so far stimulate plant growth, abscisic acid **3.30** modulates several aspects of dormancy. Indeed its early name was dormin. Abscisic acid is formed in the autumn and brings about leaf and fruit fall. It is also formed by plants under water stress and can lead to the loss of leaves in the height of summer, protecting the plant against water loss. In many ways it is an antagonist of the plant growth hormones. Although abscisic acid has a sesquiterpenoid structure, it may be formed in plants by the biodegradation of the carotenoid violaxanthin via a xanthoxin. The isoprene units of the carotenoids that are biosynthesized in the chloroplasts are derived by the deoxyxylulose pathway. Abscisic acid in plants is formed by this pathway. Abscisic acid is also formed by some fungi that are plant pathogens such as *Cercospora rosicola*. These fungi use the mevalonate pathway to form abscisic acid directly from a sesquiterpenoid precursor farnesyl pyrophosphate. Fungal and plant abscisic acids appear to be biosynthesized by different routes.

Ethylene is an unusual hormone and plays an important role in the ripening of fruit. It is formed in ripe fruit and induces ripening in others. Green tomatoes may be ripened under a sheet of paper to trap the ethylene, provided that there is a ripe tomato or even a ripe banana present to provide the ethylene. Ethylene is biosynthesized from the amino acid 1-aminocyclopropanecarboxylic acid **3.31**. The production of ethylene can affect the storage of fruit. Apples that are producing only a low concentration of ethylene (less than 0.1 ppm) will store quite well in a cold area, whilst those that are producing more ethylene do not store well. The inhibition of ethylene biosynthesis by compounds such as 2-amino-4-aminoethoxy-*trans*-3-butenoic acid (aminoethoxyvinylglycine, AVG) or the function of ethylene by the gas 1-methylcyclopropene (MCP) delays senescence and extends the storage life of fruit. Some of the 'out-of-season' fruit on the supermarket shelves has been preserved by the action of MCP.

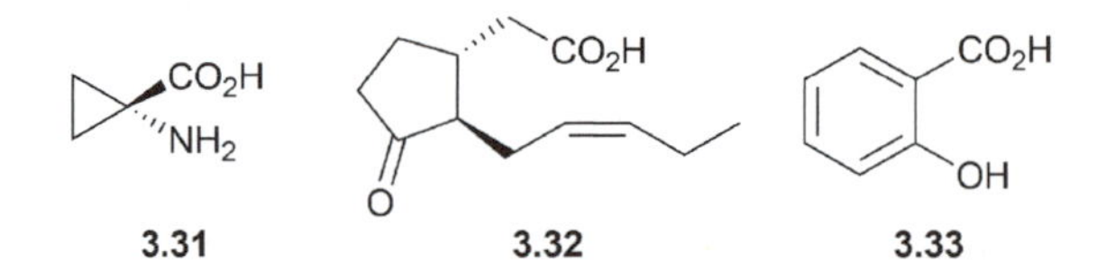

A number of other compounds play a signalling role in plant chemistry. Jasmonic acid **3.32** is produced under stress and in low concentrations. It stimulates the defensive system of plants. Application of the methyl ester of jasmonic acid induces the formation of tubers in potatoes. It is related to the natural product tuberonic acid. Jasmonic acid also induces ethylene synthesis and it can inhibit seed germination. The cyclopentenone *cis*-jasmone is produced by decarboxylation of jasmonic acid and is a well-established component of plant volatiles. It is released when a plant is damaged. Whereas it has a repellent effect on some insects such as aphids, it acts as an attractant to their predators such as ladybirds. Along with other leaf volatiles including *cis*-hex-3-enal and methyl salicylate, it also has a signalling effect on neighbouring plants, priming them against impending herbivore attack. Salicylic acid **3.33** also influences the growth of flowers, buds and roots. When it is added to the water in a vase containing cut flowers, it can delay their withering.

Thus there is a variety of natural products of widespread occurrence in plants that play an important role in general plant biochemistry. Factors that contribute to their formation underpin the use of many chemicals in the garden.

CHAPTER 4

Garden Soils

The soil represents the interface between inorganic and organic chemistry in the garden. The soil is a mixture of the weathered bedrock, and material deposited by glaciation and by rivers and from the atmosphere. It contains the decaying remains of plants and animals and a wealth of living organisms. The soil in which the plant grows provides much more than just an anchor to hold the plant in place.

In the previous chapters we have seen the role that various macromolecules and natural products play in the development of the plant. The nitrogen, phosphorus, sulfur and metal ion content of these together with water come from the soil. Much of what can take place above the ground is determined by what is available below the ground. The macroscopic properties, the mineral content and the organic humus in the soil are important in the garden in determining the health and development of the root system of plants. Water, metal ions, phosphate and other nitrogen containing compounds are transported in the xylem to the upper parts of the plant. In turn, sugars and other compounds are returned to the roots in the phloem. The roots of plants can also act as carbohydrate storage organs, a feature which is exploited in foodstuffs.

4.1 THE MINERAL STRUCTURE OF THE SOIL

Soils are broadly classified in terms of their dominant particle size. An approximate division is that particles that are greater than 2 mm in diameter form gravel and stones. Those particles between 0.1 mm and 2 mm are sand, whilst those that are between 0.005 mm and 0.1 mm are silt and those that are less than 0.005 mm are clay. It is not just the particle size which is important but also their mineral composition and surface area.

The pores within the soil are determined by the particle size. The larger transmission pores ($>50\,\mu m$ in diameter) allow for water and root penetration. These larger pores are associated with lighter sandy soils and allow fungal mycelium to spread in these soils. Narrower pores (0.2–50 μm in diameter) store water for plant use but the smallest pores ($< 0.2\,\mu m$ in diameter) retain water in a hydrogen bonded network. The pores within the soil not only affect water and thus metal ion availability but also aeration. Many of the factors which affect chromatography in the laboratory are replicated in the soil.

Whereas sand is mainly quartz or silica, the clays are the weathered products of silicate minerals. The clays contain hydrated aluminium silicates with up to 50% SiO_2 and 30-40% Al_2O_3. There are smaller amounts of Fe_2O_3, TiO_2 and hydrated CaO and MgO. These variations produce the many varieties of clay that are found in this country. Clays contain various distinct minerals such as kaolinite and montmorillonite.

The water content of clays means that these small particles bind together and are classified as heavy. In the spring they take longer to warm up and when they dry out they form a crust and there is contraction. The effect on a lawn or even worse on house foundations can be very serious. Some clays have a colloidal fraction which can be coagulated by the addition of lime. The importance of an autumn dig to allow the winter frost to penetrate and cause this interstitial water to freeze and break up the clods is well known.

The underlying materials that contribute to the soil may be igneous rocks such as granite, metamorphic rocks such as slates, or sedimentary rocks such as chalk or limestones. As the molten magma crystallized to form the igneous rocks, the mineral content changed from olivine, a magnesium silicate, and various feldspars containing aluminium silicates together with sodium, potassium, calcium and barium oxides, to iron rich minerals such as pyrites, magnetite, chromite and ilmenite. Finally, rocks with a higher silica content were formed. Rocks with less than 45–55% silica are known as basic rocks, whilst those with greater than 65% silica, which often contain free quartz, are acidic igneous rocks. Whereas the igneous rocks weather to give soils containing quartz and mica, the sedimentary rocks give calcite or dolomite (a mixture of magnesium and calcium carbonates).

In the UK the parent material of the soil has often been moved some distance by the extensive glaciation that once affected the country. This soil movement alters the nature of the soil and its stone and flint content. Water erosion reduces the alkali and alkaline earth metal content of the soils.

The structure of the silicates determines many of the properties of the resultant rock and hence the derived soil. The dominant feature of silicate minerals is the tetrahedral SiO_4^{4-} unit. In the simple olivine materials this is associated with two divalent cations such as Mg^{2+}, Fe^{2+} and Ca^{2+}. However if there are shared oxygens between SiO_4 units, a silicate chain, sheet or three-dimensional network can develop. These bind metal ions, or in some instances there is replacement of the silicon by aluminium. Aluminium hydroxide and magnesium hydroxide can also form sheet structures.

Clay minerals such as kaolin and montmorillonite contain layers of tetrahedra and sheets sometimes with cations or water molecules in the interlayer spacing. This gives rise to variable cation exchange properties and to a change in shape between the hydrated and dry materials.

4.2 THE ORGANIC CONTENT OF THE SOIL

The degradation of plant matter leads to the accumulation of humus in the soil. This organic matter serves to bind the inorganic particles together. The black colloidal powder, known as humus, is divided into alkali-soluble and insoluble fractions, the humic acids and the humin. The humic acids are high molecular weight polymeric polyhydroxy-phenolic acids that are derived, for example, from caffeic acid **4.1** and 3, 4-dihydroxybenzoic acid **4.2**. They have a variable number of poly-saccharide and protein units attached to them. A typical analysis might be 56–60% carbon, 4.5–5.8% hydrogen, 2–4% nitrogen and 33–35% oxygen. Whilst the cellulose of plant material decomposes relatively rapidly, the lignin is more resistant and this provides the basis of the humic acids. The presence of the hydroxyl and carboxyl groups provide ample opportunity for complexing metal ions particularly iron and also lead, copper, nickel and zinc. It has been suggested that humic acids may hold up to 10% of their weight as metal ions. The combination of cation exchange sites in clay minerals and the metal ion binding properties of the soil organic matter leads to the concept of a cation exchange capacity. This determines the availability of some essential nutrients, including the ammonium ion, for the developing plant.

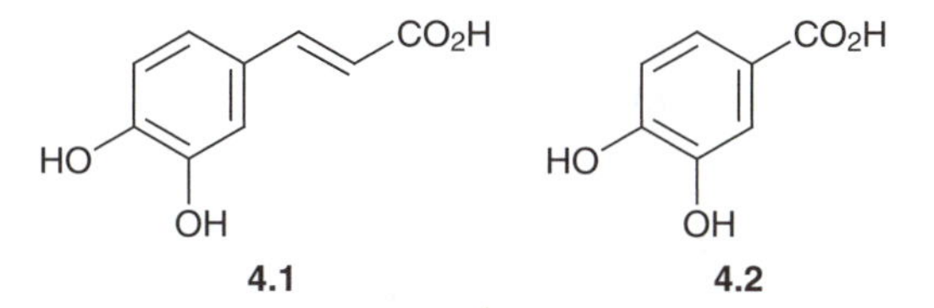

The lower molecular weight fulvic acids are more water soluble and they sometimes give a brownish colour to surface water. A normal concentration in surface water might be 1–5 ppm. A typical analysis gives a composition of 42–50% carbon, 4–6% hydrogen, 1–2% nitrogen and 45–47% oxygen. Compared to the humic acids, the fulvic acids have a higher oxygen content. The fulvic acids not only complex metal ions but may also hold phosphate units. The binding of phosphates by organic matter in the soil can seriously affect the availability of phosphate for plant growth. Inositol pentaphosphates and hexaphosphates, such as phytic acid **4.3**, can account for a substantial proportion of the total soil phosphorus. These originate in plants and are only slowly biodegraded.

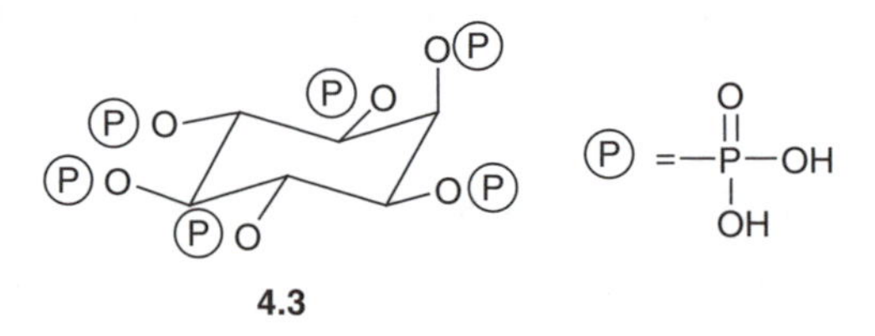

4.3

As the roots of a plant penetrate the soil, they produce a polysaccharide exudate which facilitates the uptake of water and metal ions by the root. The development of this layer is sometimes apparent around germinating seeds. It holds soil to the root system and plays a part at the interface between the soil and the plant. Care has to be taken not to disturb this in planting out seedlings. Micro-organisms in the soil also produce a polysaccharide exudate which plays an important part in the transport of nutrients. Mycorrhizal fungi associated with the root systems of plants play an important role in the mobilization of nutrients for their host plant.

The primary alcohols of the sugars in the polysaccharides that constitute the pectin of fruit are oxidized to a carboxylic acid. Biodegradation of this material from fallen fruit, and other sources of polysaccharides such as seaweed and some kitchen waste, gives a polyuronic acid which can modify the way in which clay particles bind together. Bacterial polysaccharides also have a marked aggregating effect on soil particles and thus play an important role in soil structure formation.

4.3 NUTRIENTS FROM THE SOIL

We have already seen the role of metal ions such as those of magnesium and iron in the coenzymes that mediate photosynthesis and oxidative processes. Phosphates play an important part in energy transfer as ATP, as constituents of the nucleic acids and in various biosynthetic processes.

Compounds containing nitrogen such as the amino acids and various heterocyclic bases are ubiquitous. Sulfur is a component of the amino acid cysteine and of a number of more specialized natural products. Other metal ions such as those of potassium, calcium, magnesium, molybdenum, zinc and copper also play a role in various cellular processes. Trace amounts of boron, selenium and cobalt are also required by plants. Cobalt is the central metal ion in methyl and adenosylcobalamin. These coenzymes participate in some methylation and rearrangement reactions.

The soil provides the source of these elements. Nitrogen fixing bacteria such as *Azotobacter* and the *Rhizobium* species associated with root nodules on legumes, convert atmospheric dinitrogen to ammonia. *Nitrosomonas* and *Nitrobacter* species convert ammonia to nitrites and nitrates. The decay of plant material also provides a source of nitrogen and ammonia.

Soil bacteria and fungi, by incorporating nitrogen into their structure, immobilize the nitrogen making it unavailable to plants. On the other hand, by degrading plant material, the nitrogen can be released by a process known as mineralization.

4.4 THE ROLE OF pH

The availability of metal ions for uptake by plants is particularly dependent on the pH of the soil. This is an extremely important parameter in determining the availability of nutrients for the developing root system. In an alkaline soil, iron, zinc and manganese are held as their less soluble hydroxides, whilst they are more readily available as salts from an acidic soil. These factors influence the growth of plants such as rhododendrons and azaleas. The iron deficiency can be rectified by the use of an iron–EDTA complex, Sequestrine®. An optimum pH for the growth of plants is 6.5 although some plants prefer a more acidic soil. Many bacteria grow at a pH nearer to 6.5, whilst fungi prefer a more acidic soil, an aspect of pH which can also have an effect on the development of plants.

The garden soil may also be affected by the previous use of the land. With more use of 'brown field' sites, the contamination of soil with industrial residues can present a problem. This may affect the soil pH and metal content. It is not all that long ago since the common method of disposing of industrial waste was to bury it. Buried iron, paint and electrical components can make a significant contribution to soil trace metal composition. The construction of a house can bring subsoil to the surface, whilst the presence of drains and 'soakaways', neighbouring trees and paths can produce local variations in drainage and fertility.

Some plants have the ability to concentrate toxic metals and are used in 'bioremediation'. Other plants, such as campion (*Silene vulgaris*), which is common in roadside verges, protect themselves from toxic heavy metals by complexing them with phytochelatins. These are small cysteine-rich peptides of the general structure $(\gamma\text{-glu-cys})_n$-gly. Where n is 2–11. They have a formal similarity to glutathione which is in fact a substrate for phytochelatin synthase. However, if the plants are used as foodstuffs, human toxicity may arise. Sewage sludge is sometimes used as a soil conditioner but it can contain elevated levels of zinc, copper, cadmium, lead, chromium and other elements arising from industrial waste. Edible crops, such as cabbage and onions, grown on soil to which 10% sludge had been added, were found to contain substantially higher levels of cadmium. Problems of cadmium poisoning associated with building on former mine workings, have occurred as a result. In the days of coal fires, there was also a questionable practice of using soot from the chimney, around onions.

Lignin degrading fungi such as *Phanerochaete chrysosporium, P. sordida* and *Trametes hirsuta* have been shown to degrade the wood preservatives pentachlorophenol and creosote in soil, whilst other organisms have been used to remove some polycyclic aromatic hydrocarbons from the soil of former gas works. The ability of plants and fungi to assimilate trace metals from the soil was demonstrated by the studies on the widespread fall-out from the Chernobyl explosion. ^{137}Caesium was detected in *Boletus* species of fungi growing many hundreds of miles distant from the explosion. Twenty years after the event, Welsh sheep are still monitored for radioactivity.

4.5 FERTILIZERS AND COMPOST

Deficiencies of essential elements in the soil can lead to problems of plant growth. Restrictions in the availability of nitrogen reduces the biosynthesis of amino acids, chlorophyll and the nucleic acids, thus affecting the growth and colour of the plant. A shortage of phosphorus restricts the growth of plants by reducing the availability of phosphate for ATP and nucleic acid formation as well as various other biosynthetic steps. A deficiency of magnesium and iron restricts the formation of chlorophyll and the haem pigments producing chlorotic plants. Magnesium is also required along with ATP in isoprenoid biosynthesis. Potassium ions are involved in water transport across membranes. A potassium deficiency is reflected in the browning of leaf tips.

Fertilizers containing ammonium nitrate or ammonium sulfate, calcium phosphate and potassium chloride or nitrate are used to rectify these deficiencies. It is worth noting that when an ammonium salt such as ammonium nitrate is used as a nitrogen source, the ammonium ion is assimilated first and the soil becomes more acidic. Urea is sometimes used as a nitrogen source. The composition of fertilizers is not always specified directly as percentage nitrogen, phosphorus or potassium but as percentage nitrogen, phosphorus pentoxide and potassium oxide although no fertilizer would contain P_2O_5 or K_2O! This reflects the analytical methods that were used to quantify these elements. Hence a typical Growmore fertilizer that is described as NPK 7:7:7 has a total nitrogen content of 7%, a phosphorus content of 3% and a potassium content of 5.8%, whilst a 4:4.5:8 Tomorite with added magnesium ions contain 4% nitrogen, 2% phosphorus, 6.6% potassium and 0.018% magnesium.

The use of the chemically much more complex manures and compost as a source of nutrients involves a balance between the requirements of bacteria and plants, particularly in terms of the availability of nitrogen. Bacteria have a lower carbon to nitrogen ratio (approximately 4:1) compared to the plant material that is being returned to the soil via the compost heap (approximately 20–30:1). Consequently a compost heap with actively growing bacteria can become depleted in nitrogen as the bacteria immobilize the nitrogen in their peptides and proteins. Indeed an effective compost heap may need feeding with a nitrogen source. The compost which is spread on the soil, whilst it may be beneficial for other reasons, can at first deplete the soil of nitrogen. The availability of nitrogen from manure can depend on the extent to which bacterial decomposition has proceeded, the straw content and the animal source. Manure derived from poultry, particularly with a high urea content, can release ammonia rapidly and in sufficient quantities to be toxic to plants.

The depth of the topsoil and its relationship to the subsoil and any underlying chalk, sand or granite can have an impact on the growth of deep-rooted shrubs and trees. Ericaceous shrubs such as rhododendrons, with a requirement for iron, cannot be grown on chalk.

4.6 MICROBIAL INTERACTIONS WITHIN THE SOIL

The soil is a dynamic biological medium in which bacteria, fungi, nematodes and various insects flourish in competition with the developing plant. The microbial metabolite geosmin **4.4** contributes to the smell of newly dug soil. The roots of many plants have mycorrhizal fungi associated with them in a symbiotic relationship. These fungi are divided into four groups: sheathing, vesicular-arbuscular, orchidaceous and ericaceous. The sheathing mycorrhizal fungi are often associated with trees such as the pines, spruce and larch. Some are Basidiomycetes and have quite conspicuous fruiting bodies. For example the fly-agaric *Amanita muscaria* is associated with birch or pine trees, whilst the truffle *Tuber melanosporum* occurs naturally in association with oak and hazel. The vesicular-arbuscular mycorrhizal fungi grow outwards from a few points on the root of their host plant. They can penetrate the roots at these points of attachment. Many orchids will not grow without their symbiont fungi. The ericaceous mycorrhizal fungi are particularly associated with the heaths and rhododendrons, facilitating their growth on poor soils. Whilst these fungi draw their sugars from their host plant, they mobilize nutrients from the soil to the benefit of the plant. They are particularly important in releasing phosphate and mobilizing iron.

The beneficial effect of fungi in releasing nutrients can be seen in the enhanced growth of grass just within a 'fairy ring' (*Marasmius oreades*) in the lawn. Further inside the ring the grass may then die. Since many bushes and trees will not flourish without their associated mycorrhizal organisms, in transplanting bushes it is helpful to ensure that the soil associated with the plant is moved as well. When shrubs or trees are planted on reclaimed brown field sites there may well be a local deficiency in available nutrients such as phosphate. The introduction of mycorrhizal organisms can remedy this by facilitating the release of phosphate. The roots of plants have been found to stimulate the germination of mycorrhizal fungi. Recently this activity has been associated with derivatives of deoxystrigol **4.5**. Soil bacteria, such as *Pseudomonas aureofaciens* and *P. putida* and some fungi, produce siderophores, metabolites which complex iron. These complexes, which can involve a hydroxamate or a catechol, have the effect of mobilizing iron, particularly Fe(III), in the rhizosphere and increasing its bioavailability to the roots of plants.

Although many soil bacteria are involved in the biodegradation of plant material, others carry out biological nitrogen fixation. The free-living *Azotobacter* species and the symbiotic *Rhizobium* species found in the root nodules of legumes such as peas and beans carry out this nitrogen fixation converting atmospheric nitrogen into ammonia. The nitrogenase enzyme has been thoroughly studied. It contains two units. The smaller unit has a molecular weight of approximately 60 000 and contains four iron atoms. The larger protein with a molecular weight of approximately 220 000 contains as many as thirty iron atoms and crucially two molybdenum or, less commonly, two vanadium atoms. The larger of the two proteins carries out the actual reduction of the nitrogen whilst the smaller is involved in the electron transfer mechanism.

A heavily compacted waterlogged soil can become anaerobic. Anaerobic bacterial decay leads to the production of methane. In a marshy area this can ignite as a 'will-o'-the-wisp'. It is possible that sufficient phosphine is also produced to ignite the gas on contact with air. This release of gas has also been a problem with houses built on brown field sites that have previously been landfill sites.

On digging the soil, the soil that is brought to the surface may lighten in colour as the darker low valency iron sulfides are oxidized. It is sometimes possible to smell the hydrogen sulfide. The 'earthy' smell of newly dug soil is due to a norsesquiterpene, geosmin **4.4**, which is produced by soil *Streptomycete* species. The presence of geosmin and 2-methylborneol in water supplies can be a problem because the human nose can detect them at very low concentrations.

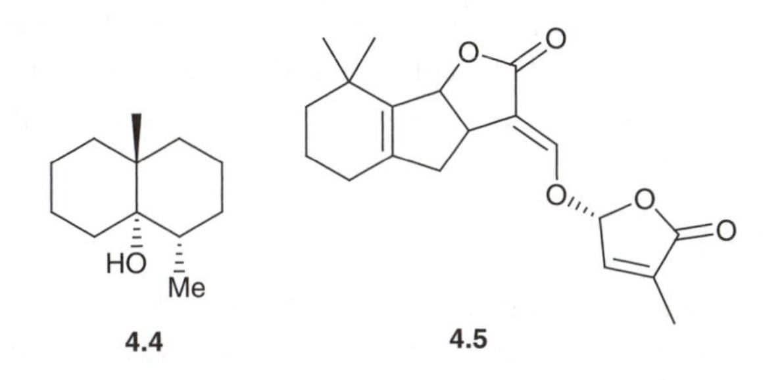

Digging the soil not only alters its structure but in enhancing aeration and improving drainage it releases various gases that can build up. It also disrupts the effect on insect trail substances and it can reduce the impact on seed germination of allelopathic agents released by plants. The effect of these compounds will be discussed in later chapters.

CHAPTER 5

The Colour and Scent of Garden Plants

Many ornamental garden plants are grown for the colour of their flowers and leaves or for the scent that they possess. The compounds that contribute to these properties form the subject of this chapter.

5.1 COLOURING MATTERS

Two major groups of compounds which contribute to the colours of garden plants are the carotenoids and the anthocyanins. Apart from the green colour of chlorophyll in the leaves, some plants also contain quinone and betalain pigments. Whereas the carotenoids are relatively non-polar molecules, the anthocyanins which contain a sugar moiety attached to the anthocyanidin aglycone, are highly polar molecules.

5.2 THE CAROTENOIDS

The carotenoids are terpenoid substances containing eight isoprene units. They owe their colour to a series of eleven or twelve conjugated double bonds. They are responsible for many of the yellow, orange and red colours in plants. Carotenoids are also found in animals, algae, fungi and bacteria. A hydrocarbon known as 'carotene' was originally isolated from carrots in 1831 by Wakenroder. In 1931 in an early demonstration of the power of chromatography, Kuhn was able to separate the carotene mixture obtained from carrots into α, β and γ isomers. The predominant isomer was β-carotene **5.1**. The structure of β-carotene was determined by Karrer and others in the 1930s. The variation in colour of the carotenoids is associated with the length of the conjugated system, the possibility of geometrical isomerism about the double bonds and the oxidation, in particular epoxidation, of the terminal double bonds.

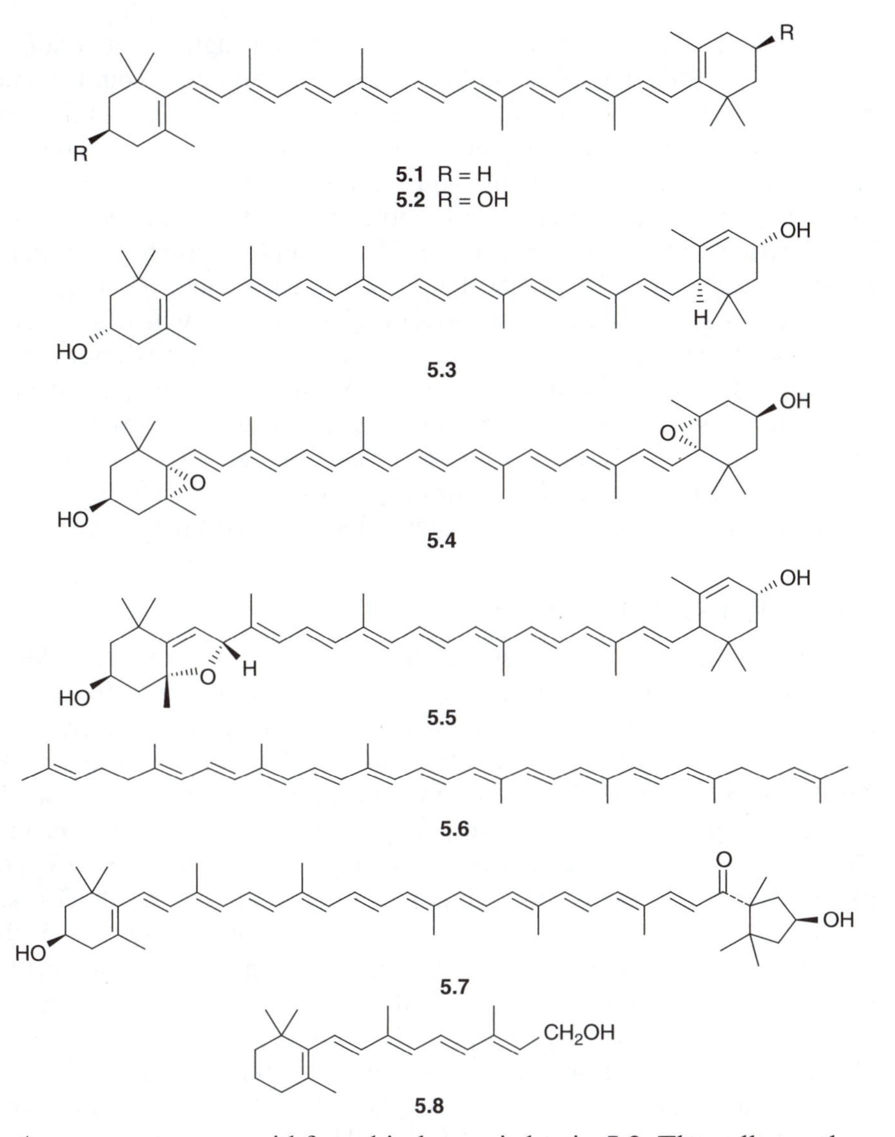

A common carotenoid found in leaves is lutein **5.3**. The yellow colour of crocuses and daffodils (zeaxanthin **5.2**) and of pansies (violaxanthin **5.4**) arises from carotenoids. Dandelions contain the carotenoids lutein 5,6-epoxide, also known as taraxanthin, and flavoxanthin **5.5**. The red colouring matter of tomatoes (approx. 20 mg/kg) is the carotenoid, lycopene **5.6**. Peppers (*Capsicum annuum*) are another source of carotenoids. Capsanthin **5.7** provides the red colour. Lutein **5.3** and the 5, 6-monoepoxides and 5,6;5′6′-diepoxides of β-carotene are found in the yellow and orange colours of the African and French Marigolds (*Tagetes erecta* and *T. patula*).

The colour of roses has been the subject of thorough investigation. It has been estimated that of the 50 000 hybrids that have been commercially produced over the last three to four centuries, about 10% are still found in various collections. Consequently their petals have been available for study and their carotenoid and anthocyanin content has been documented. Over seventy different carotenoids have been identified in roses.

The carotenoids are powerful anti-oxidants and there are health benefits to be gained from eating a diet that is rich in carotenoids. Furthermore β-carotene **5.1** is converted in the liver to vitamin A **5.8** which is essential for vision. The biodegradation of the carotenoids also affords a number of flavouring substances such as β-damascenone and β-ionone (see later).

Carotenoids which occur in marine organisms such as lobsters, do so in association with a protein. The typical carotenoid colour is only revealed when this complex is destroyed on boiling. There is some evidence for the presence of a carotenoid–protein complex in carrots.

5.3 THE ANTHOCYANINS

The anthocyanins that are found in the petals of plants are glycosides of four common anthocyanidins: pelargonidin **5.9**, cyanidin **5.10**, delphinidin **5.11** and peonidin **5.12**. The structures of many of these pigments were established by the work of Willstatter and Robinson in the 1920s. The anthocyanidins differ in the oxygenation of one ring and hence in the wavelength of their absorption maxima in the UV–visible spectrum. Typical of phenols, this absorption is very pH dependent, see **5.13**. The pH of the sap of the cells containing the anthocyanins is typically 3.7–4.2 and hence the anthocyanins exist as salts. When the pH rises to 4.5, the colourless hydrates can form and the oxygen ring can open to form a chalcone. At one time it was thought that the blue colour of flowers arose from the ionization of the anthocyanins in alkali in the plant. However it was questionable that flower sap could become sufficiently alkaline and there are now other explanations for these colours (see later). Nevertheless this aspect of the pH sensitivity of anthocyanin colours is apparent to anyone who has washed up a fruit dish that has had cooked blackcurrants in it and used soda in the washing water.

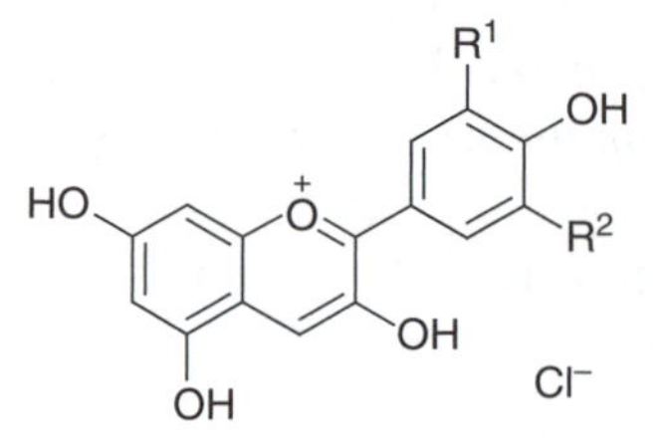

5.9 $R^1 = R^2 = H$
5.10 $R^1 = OH, R^2 = H$
5.11 $R^1 = R^2 = OH$
5.12 $R^1 = OH, R^2 = OMe$

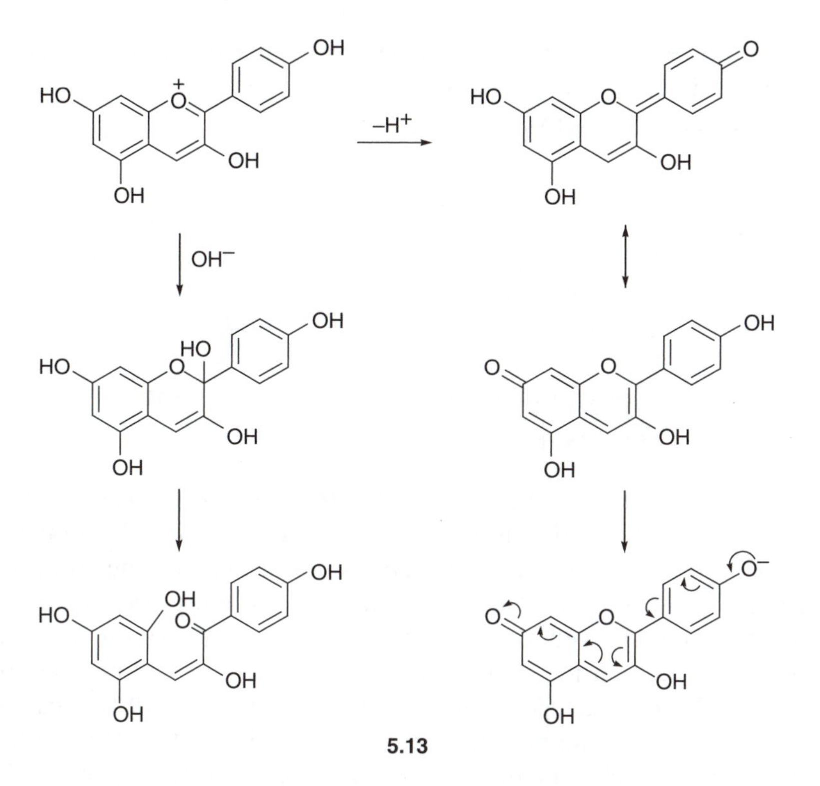

5.13

The major anthocyanidins of the rose are cyanidin and pelargonidin. The variation in colour arises from co-pigmentation with a flavonol such as kaempferol 3-*O*-β-D-glucoside **5.14**. Whereas the anthocyanidin ring system is electron-deficient, the flavones and flavonols are electron-rich and hence complexes between the two can form. In other flowers such as delphiniums, a blue colour arises from co-pigmentation with anthocyanins derived from delphinidin. The idea of co-pigmentation was first put forward by Robinson in 1931.

Anthocyanins derived from the anthocyanidin delphinidin have not been found in roses and hence a true blue rose is elusive. Investigation of a number of cultivars, for example 'Rhapsody in Blue' which display an almost blue colour, showed that the colour is based on co-pigmentation between cyanidin **5.10** and kaempferol **5.14** or quercetin **5.15** glycosides. As the petals age, the anthocyanin becomes attached through hydrogen bonding to the protein matrix in vacuoles in the petals and this produces a second absorption maximum in the 620–625 nm range, giving rise in combination with other absorption to a purple colour. Interestingly the tannins present in rose petals are based on gallic acid, 3,4,5-trihydroxybenzoic acid **5.17** and hence some of the biosynthetic apparatus exists in the rose for the production of the delphinidin substitution pattern.

The blue flowers of *Ceanothus* species, particularly *C. papillosus* (the Californian lilac) involve co-pigmentation between a delphinidin glycoside and a kaempferol glycoside.

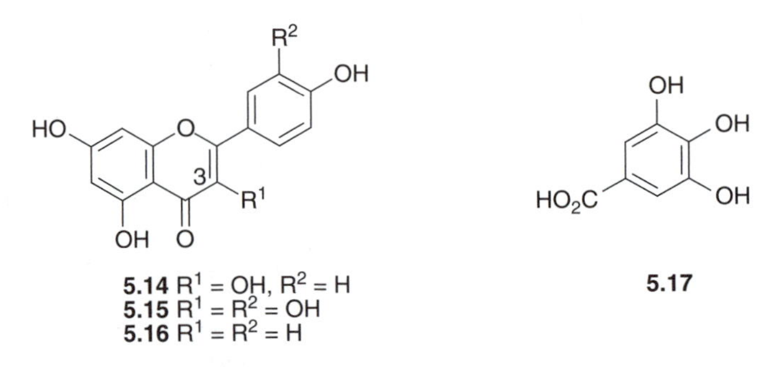

5.14 R^1 = OH, R^2 = H
5.15 R^1 = R^2 = OH
5.16 R^1 = R^2 = H

5.17

The blue flowers of the cornflower (*Centaurea cyanus*) illustrate another aspect of the development of colour, the role of metal ions. The blue pigment, protocyanin, has been obtained crystalline and shown to contain iron, magnesium and calcium ions. An X-ray crystal structure of this complex showed that it contained six anthocyanin molecules (cyanidin 3,5-diglucoside), six flavone molecules (the 7-*O*-methyl ether of apigenin **5.16**), one ferric iron, one magnesium ion and two calcium ions. The iron and magnesium are each co-ordinated to three anthocyanins, whilst each calcium is co-ordinated to three flavones. The anthocyanins and the flavones are stacked and the bond lengths suggest that the anthocyanin is in the quinone-methide form. The blue colour of the flowers of *Salvia patens* arises from a complex between six molecules of delphinidin glycoside, malonylawobanin, six molecules of a flavone, apigenin 7,4′-di-*O*-β-D-glucoside, together with two magnesium ions. If these compounds are mixed together in aqueous solution in the correct ratio, they form a blue chiral metallo-anthocyanin complex.

Aluminium is another ion which is involved in complexes of this type. The blue sepal colour of *Hydrangea macrophylla* has been associated with the presence of a metal-complex pigment which consists of delphinidin 3-glucoside, the co-pigments 5-*O*-caffeoylquinic acid or 5-*O*-*p*-coumaroylquinic acid and aluminium ions. The variation in the colour of hydrangea sepals from pink and mauve to blue has been correlated with the aluminium content of soils. Feeding a hydrangea with aluminium sulfate can restore a deep blue colour to a plant that has begun to revert to purple or even pink. On the other hand, tying up the metal salts with lime gives pink colours. Many more blue flowers require an acidic ericaceous soil with increased metal ion availability.

The diversity of colours in antirrhinum (snapdragon, *Antirrhinum majus*) has been investigated. The 3-rutinoside of pelargonidin **5.9** contributes to the pink flowers, whilst the 3-rutinoside of cyanidin **5.10** with the extra hydroxyl group produces the magenta or crimson flowers. The yellow colours are associated with the flavones, quercetin **5.15** as its 3-glucoside and apigenin **5.16** as its 7-glucuronide. The genetics of these colour variations has been studied.

Glycosides of flavones, including quercetin **5.15**, are the major yellow pigments of some *Primula* species, *e.g.* the primrose, where they occur with carotenoids. The intense red-scarlet colour of *Salvia splendens* and of geraniums are derived from glycosides of pelargonidin **5.9**.

Apart from attracting insects to flowers for pollination purposes, both anthocyanins and carotenoids absorb ultraviolet radiation and act as ultraviolet screens. White flowers still contain pigments of these types but which do not absorb in the visible region of the spectrum. These pigments are also powerful anti-oxidants preventing free radical damage to the plant.

The anti-oxidant activity of anthocyanins confers beneficial health effects for man. The oxidative modification of low density lipoprotein has been implicated in the development of artherosclerosis and coronary heart disease. Dietary anti-oxidants including anthocyanidins such as delphinidin with a high radical scavenging activity protect the low density lipoprotein from oxidation. The value of these compounds in various fruits and vegetables is discussed later.

Although we associate the green colour of chlorophyll with the leaves of plants, anthocyanins and carotenoids are also present in the leaves but their colours are masked. Nevertheless they exert a protective role in the leaf. The carotenoids that are present are typically β-carotene **5.1** and lutein **5.3**. As colouring matters, these carotenoids and anthocyanins come into their own in the autumn. The onset of shorter days and lower temperatures brings about changes in the structure and biochemistry of the leaf. Abscission cells grow across the join between the leaf stalk and the stem, reducing the flow of nutrient to the leaf and the flow of sugars from the leaf to the plant. The synthesis of chlorophyll ceases and the existing chlorophyll undergoes biodegradation. The chlorophyll is degraded prior to the other leaf pigments. Sugars accumulate in the leaf and the carotenoids and anthocyanins remain. The carotenoids are often more highly oxidized to give the 'autumn xanthophylls' and hence the leaves take on the yellows, reds and browns that afford their autumnal colours. The presence of sugars provides nutrient for degradative micro-organisms after leaf fall.

A number of plants are grown for their decorative leaves. Although, in some of these, anthocyanin and carotenoid pigments predominate (*e.g.* *Begonia* sp.) in others the pigments belong to different classes of natural products. *Coleus* species produce a range of yellow to red diterpenoid quinones known as the coleons, *e.g.* coleon A **5.18**. These are found in glands on the underside of the leaves. *Coleus blumei* is a greenhouse plant which is often used to decorate the stage for ceremonial occasions. Several of the diterpenoid pigments have an interesting biological activity as cytotoxic agents, whilst they can also, like many quinones, be responsible for contact dermatitis. One Indian member of the genus, *C. forskolii*, which is also used as an indoor ornamental plant, produces a diterpenoid forskolin **5.19** which has attracted interest because it lowers blood pressure and stimulates the adenylate cyclase system. The red betalain pigments, *e.g.* betanin **5.20**, are discussed later in the section on beetroots in the chapter on vegetables.

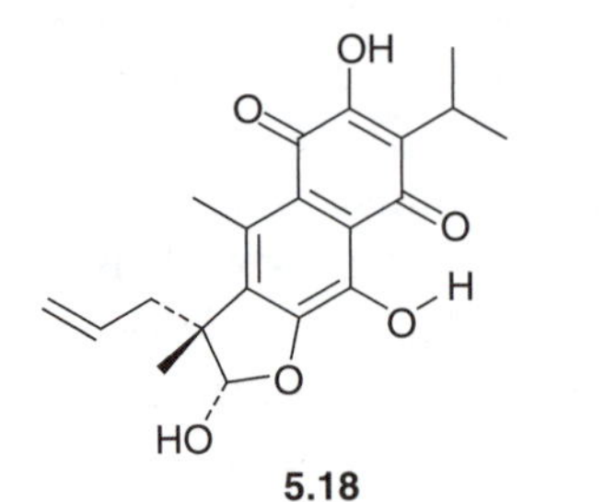

5.18

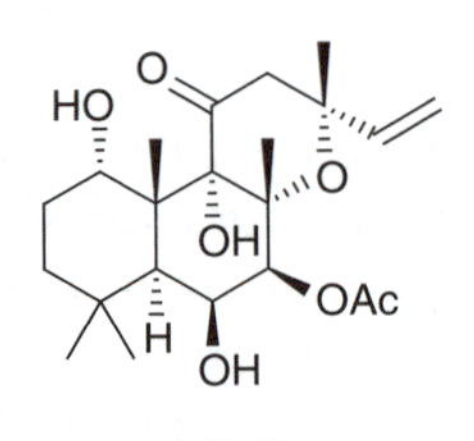

5.19

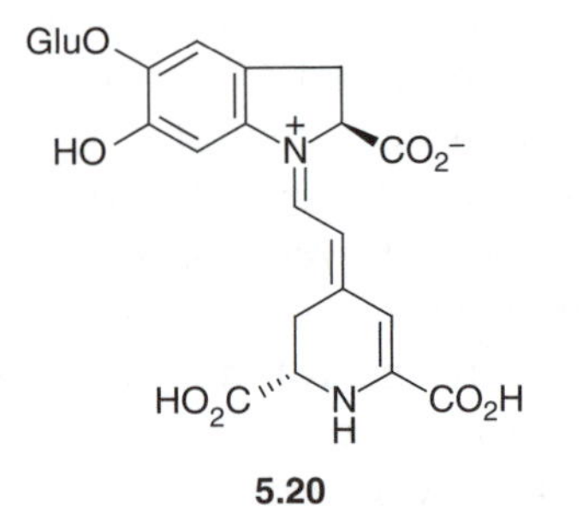

5.20

5.4 NATURAL PIGMENTS

The extraction of colouring matters from garden plants is something which most chemists are acquainted with whether it be the extraction and chromatography of chlorophyll from spinach or grass or the use of anthocyanins as pH indicators. The Easter custom of boiling an egg in an onion skin to stain the shell yellow depends on the quercetin **5.15** in the onion skin.

A number of plants that are grown in the garden contain colouring matters that were used prior to the development of the nineteenth century dyestuff industry. *Berberis* wood contains the yellow alkaloid, berberine **5.21**. Yellow pigments were also obtained from broom (*Cytisus scoparius*), goldenrod (*Solidago canadensis*) and the meadow saffron (*Crocus sativus*). However many of these pigments were too soluble to remain attached to the cloth. The cloth was therefore treated with a mordant such as an iron or aluminium salt, or a tannin such as that from oak marble gall, to hold the pigment in place.

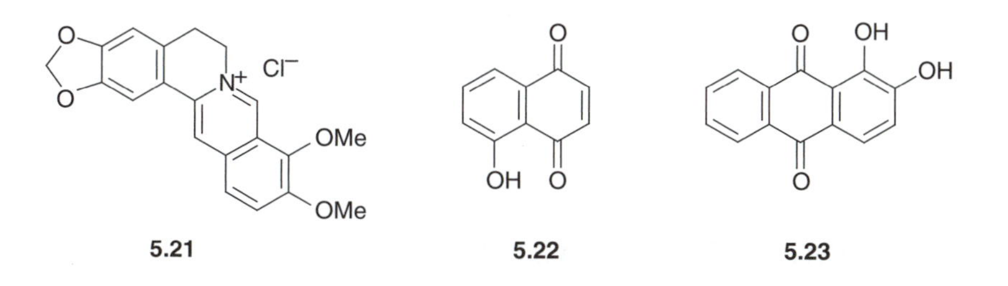

The use of natural colouring matters in food has raised the problems of stability and permanence of colour. Thus the anthocyanins that contribute to the red colour of cherries can be used to impart a red colour to other foods. These can be stabilized with phosphoric acid and attached to cyclodextrins. Although not a garden plant, the ripe seeds of the tropical Annatto tree, *Bixa orellana*, provide the degraded carotenoid, bixin, which is used to impart an orange to red colour to some cheeses.

A number of wood pigments which have attracted interest are quinones. Juglone **5.22**, a naphthoquinone from walnut and noted for its allelopathic properties, has been used as a red pigment. Other quinones such as alizarin **5.23** are obtained from madder root, *Rubia tinctorum*.

A plant which has attracted interest, and is sometimes grown as a garden plant, is Woad, *Isatis tinctoria*. There was a significant mediaeval European industry involved in producing the blue dye indigo. Woad leaves contain two indoxyl derivatives, indican (indoxyl-3-*O*-β-D-glucoside **5.24**) and a minor component isatin B (indoxyl-3-(5-ketogluconate)). In the production of indigo **5.26**, the leaves were crushed to a pulp, kneaded into balls and dried. When required, the balls were dampened and allowed to ferment. This released indoxyl, some of which underwent aerial oxidation to isatin **5.25**. The condensation of the indoxyl and isatin gave indigo **5.26** which would be used in dyeing wool. Typical yields of indigo were about 3 mg/g fresh weight of the leaves. The deep red betalain pigment of beetroot and the light red carotenoid pigments of tomatoes and carrots have also been extracted for use as separate food pigments.

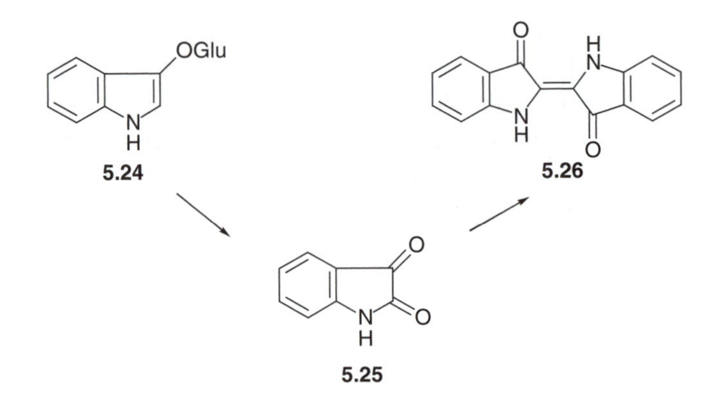

5.5 FLORAL AND LEAF SCENTS

The attractive odours of leaves and flowers have stimulated many investigations. The human nose has about 350 different odorant receptors. Each receptor can recognize a number of odorants, whilst each odorant can bind to several receptors with a varying efficiency. Odorants are recognized by the different combinations of these receptors that they stimulate. Since most natural scents are themselves a combination of odorants, there is an almost limitless range of plant scents that can be distinguished by the human nose. The volatile compounds that contribute to these scents are monoterpenes and some simple aromatic and aliphatic alcohols and esters.

The advent of gas chromatography linked to mass spectrometry has meant that many complex mixtures can now be analysed and in some cases differentiation has been made between the compounds produced by the leaves, the flowers and the pollen. It is apparent that although some compounds predominate in particular species, there are substantial variations in the detailed content even between different cultivars of the same species. This may be coupled with seasonal and diurnal variations, factors with which most gardeners are well aware.

The volatile products of flowers act as pollination attractants for insects. Those plants such as 'night-scented' stocks which rely on moths for pollination produce more volatiles in the evening.

The essential oil of roses is dominated by geraniol **5.27** and its *cis* isomer, nerol **5.28**, together with the (*R*)-isomer of linalool **5.29**. The (*S*)-enantiomer of linalool is found in coriander oil. Geometrical isomerism and chirality both affect our perception of odour. However minor components can make a significant contribution to the scent, and a cyclic ether, rose oxide **5.30**, has attracted considerable interest in this context. Another cyclic ether, lilac alcohol **5.31**, which is formed from linalool, contributes to the scent of the lilac, *Syringa vulgaris*.

The difference between the volatiles of the pollen and the whole flower has been investigated with the Japanese rose *Rosa rugosa*. Whereas the alcohols, geraniol **5.27** and nerol **5.28**, are found in the flowers, their more lipophilic acetate esters are found along with some long chain hydrocarbons, together with aldehydes and esters based on tetradecanal and tetradecyl acetate, in the pollen. The solubility of the volatile esters in these lipids means that their odour persists for a longer period of time. In a number of cases the monoterpenoids are also stored in the plant as glycosides, for example, geraniol β-D-glucoside.

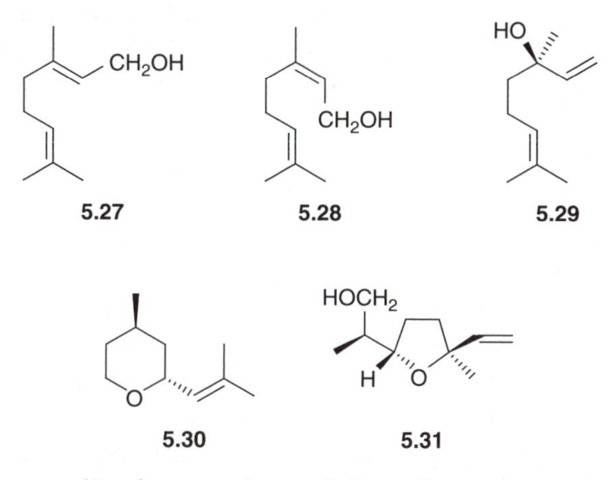

The sweet pea (*Lathyrus odoratus*) is a favourite ornamental plant which is grown for the sake of its cut flowers. One analysis of the volatiles identified forty-eight compounds and showed that the scent comprised about 40% monoterpene hydrocarbons, predominantly (*E*)-β-ocimene **5.32**, and about 30% of the alcohols, geraniol **5.27**, nerol **5.28** and linalool **5.29**. There were smaller amounts of phenylacetaldehyde and the sesquiterpenes bergamotene **5.33** and caryophyllene **5.34**. The flowers of jasmin (*Jasminum officinale*) contain benzyl acetate, indole, *cis*-jasmone **5.35**, methyl jasmonate **5.36**, methyl anthranilate **5.37**, benzyl alcohol, linalool **5.29** and farnesol **5.38** together with many minor components. This illustrates the complexity of the scents of many flowers.

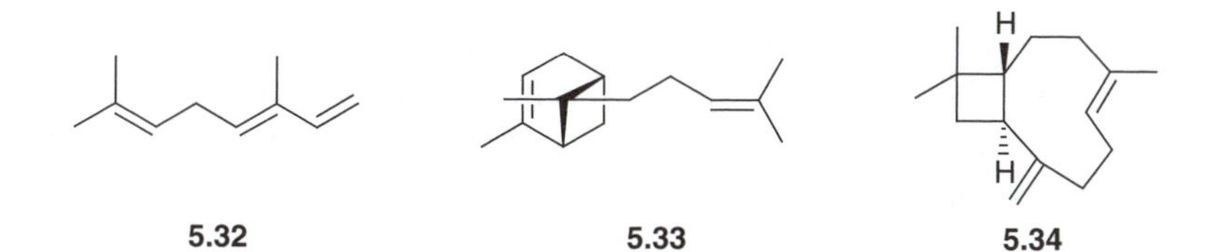

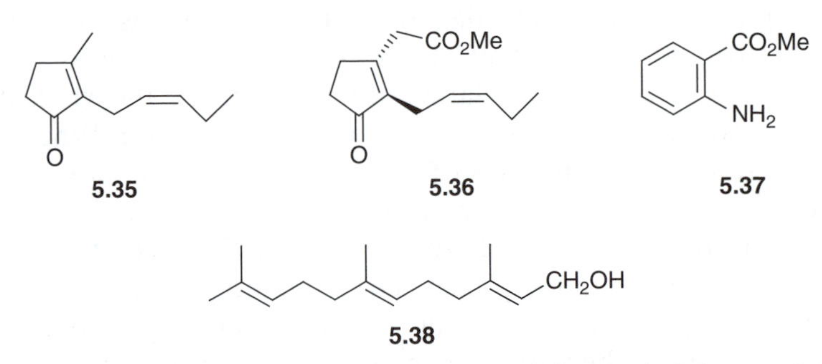

5.35 **5.36** **5.37**

5.38

The major odiferous compound of many cut leaves is 3-hexen-1-ol **5.39**. The sharper smelling aldehyde and its isomer, *trans*-2-hexenal, are found in the odour of freshly mown grass. These volatile compounds are formed by oxidation of the fatty acid linolenic acid **5.40** in the plastids. The aldehydes provide defence against microbial attack to the damaged leaf.

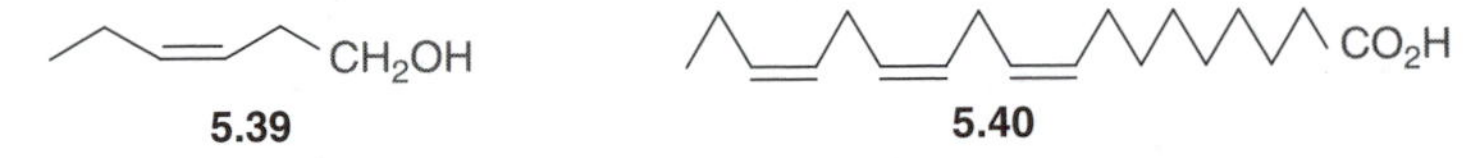

5.39 **5.40**

Some plants are grown for the scent of their leaves, particularly lavender (*Lavandula angustifolia*). Again many compounds have been detected in the leaf glands but the principal compounds are linalool **5.29** and linalyl acetate with smaller amounts of β-ocimene **5.32**, 1,8-cineole **5.41** and camphor **5.42**. An usual irregular monoterpenoid is lavandulol **5.43** which is found together with its acetate in lavender. Santolina (*Santolina chamaecyparis*) produces a similar monoterpenoid, artemisia ketone **5.44**. This compound has an irregular monoterpenoid carbon skeleton in which the tail of one isoprene unit is linked to the central carbon atom of the other.

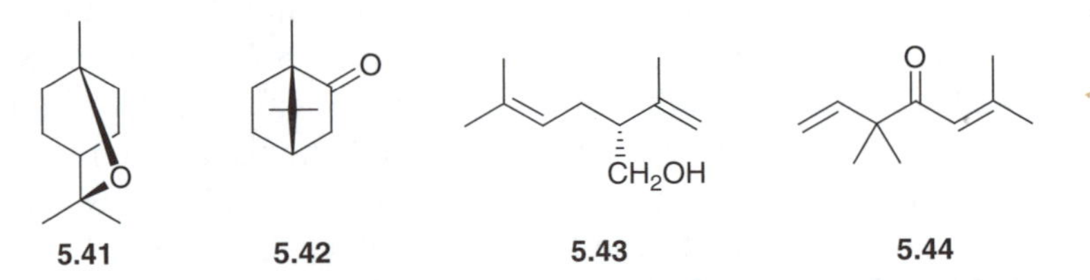

5.41 **5.42** **5.43** **5.44**

The mint family of herbs are grown for their culinary and flavouring properties which are associated with their monoterpenoid constituents. Although the major volatile components are related by possessing a *p*-menthane skeleton, they differ in their oxidation level and stereochemistry. This has a significant impact on their olefactory properties. The common spearmint (*Mentha spicata*) produces (–)-carvone (4*R*) **5.45** as a major component. This has a typical mint odour. Its enantiomer (+)-carvone (4*S*) is a major component of caraway oil, obtained from the seeds of caraway, *Carum carvi* (Apiaceae), and this has a different 'caraway' odour. The round-leafed mint, *Mentha rotundifolia*, has a slightly different odour to the spearmint and contains more pulegone **5.46** and isopulegol. Pulegone **5.46** is produced by *Mentha pulegium* and is the main component of oil of pennyroyal. This oil is an insect repellent. The peppermint, *Mentha x piperita*, is a hybrid of *M. spicata* and *M. aquatica*. It produces a mixture of menthol **5.47**, menthone, menthyl acetate, the stereoisomer neomenthol and menthofuran **5.48**. Although menthofuran **5.48** and the lactone mintlactone **5.49** are relatively minor constituents, they are quite significant sensory components. Steam distillation of the fresh leaves of the field mint or corn mint, *M. arvensis*, gives an oil from which menthol can be obtained by crystallization. Menthone and menthyl acetate are also present.

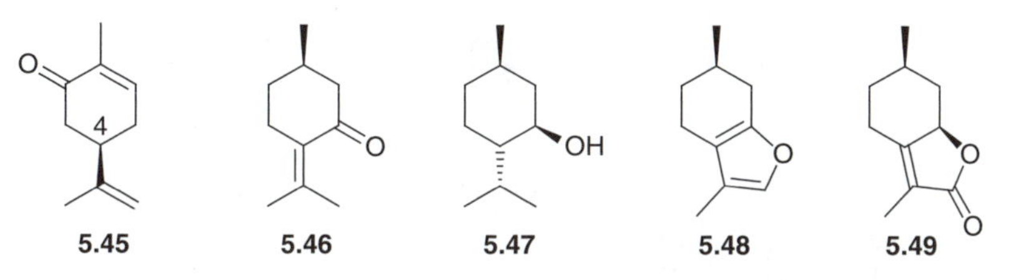

Several other common garden culinary herbs belong to the same Lamiaceae (Labiatae) family. The oil from the leaves of rosemary, *Rosmarinus officinalis*, contains up to 40% 1,8-cineole **5.41**, a comparable amount of α-pinene **5.53** and its double bond isomer β-pinene, together with some camphor **5.42**. These monoterpenes not only contribute to its flavour but also confer medicinal properties some of which are described in the next chapter. The antiseptic properties were known to our forebears who used these herbs in the preparation and storage of food. These properties have also been used by other mammals and birds. For example it has been reported that birds such as starlings when reusing an old nest will line it with herbal plant material to reduce the presence of parasites. Rosemary oil is quite effective against insect and mite pests. Its constituents have been investigated in the control of the spider mite, *Tetranyctus urticae*, which is a serious pest in greenhouses. A number of other terpenoid essential oils have insecticidal properties.

The aromatic properties of thyme (*Thymus vulgaris*) have figured in literature as in the Shakespearean phrase 'bank whereon the wild thyme blows'. The oil of thyme has a high phenol content including thymol **5.50** and carvacrol **5.51** which give it a powerful antiseptic action. Oil of thyme was used in herbal medicines as an expectorant in the treatment of bronchitis. These antiseptic properties of thymol are used in some fillings in dentistry, in mouthwashes and in throat pastilles (glycerine thymol compound BPC), whilst carvacrol is a constituent of the oil that is used in several 'vaporisers' to alleviate the congestion of a cold. Clary sage oil (*Salvia officinalis*) contains linalyl acetate but some other sages have different monoterpene components. The oil obtained from basil (*Ocimum basilicum*) contains linalool **5.29**, citronellol **5.54** and limonene **5.55** together with the C_6–C_3 aromatic ethers, methyl chavicol (estragole) **5.56** and eugenol **5.57**. Marjoram oil from *Origanum vulgare* contains 1-terpinen-4-ol **5.58**, sabinene **5.59** and *p*-cymene **5.52**. The calming effect of a combination of monoterpenoids including citral, geraniol, nerol and citronellol play a role in herbal teas prepared from plants such as lemon balm (*Melissa officinalis*) and in preparations used in aromatherapy.

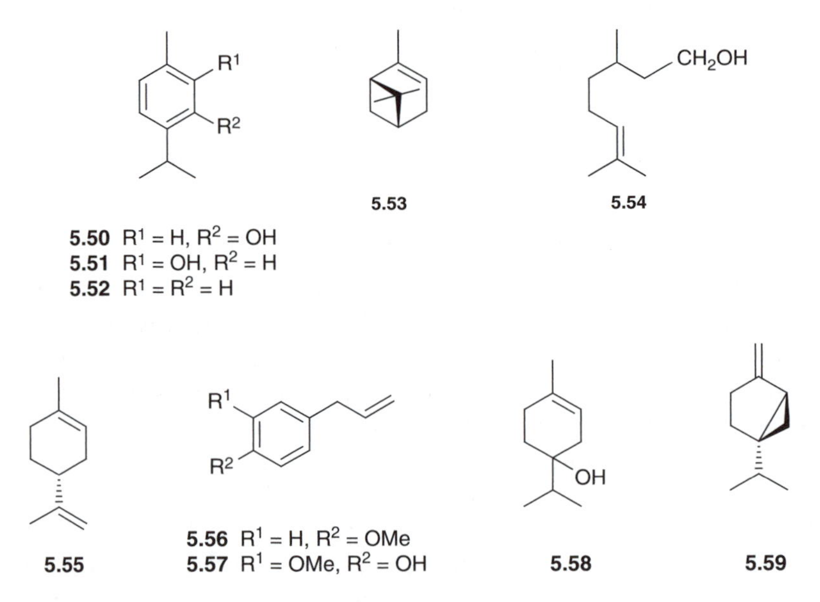

Not all aromatic culinary herbs belong to the Lamiaceae. Some such as dill (*Anethum graveolens*) and fennel (*Foeniculum vulgare*) are members of the Apiaceae. The major olefactory components of fennel oil are the phenylpropanoids anethole **5.60** and its double bond isomer estragole, together with smaller amounts of the monoterpenoid fenchone **5.61**. Dill oil contains limonene **5.55**, carvone **5.45** and dill ether **5.62**. The phenylpropanoid, dill apiole **5.63**, is also present. The leaves of parsley (*Petroselinum crispum*) which are used in parsley sauce, contain a range of monoterpene hydrocarbons including limonene **5.55**, β-phellandrene, the α- and β-pinenes **5.53** and *p*-mentha-1,3,8-triene which are responsible for the organoleptic properties. The seed oil contains rather less of the monoterpenoid constituents and more phenylpropanoids such as apiole **5.64** and 1-allyl-2,3,4,5-tetramethoxybenzene. It also contains a small amount of elemicin **5.65** and myristicin **5.66** which have a nutmeg odour.

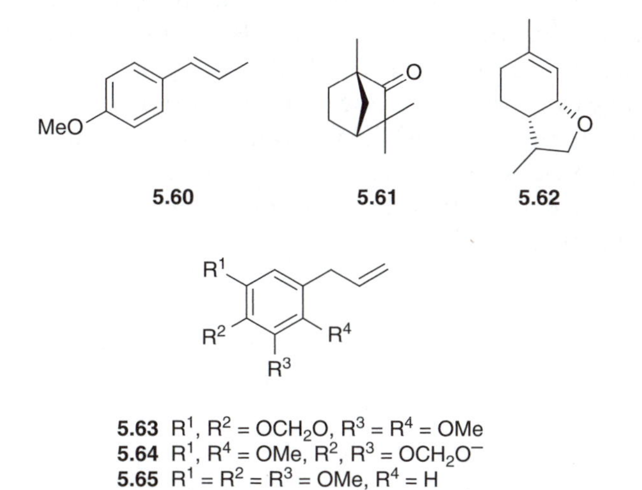

Some plants have an odour which is attractive to animals. For example the common catnip or catmint (*Nepeta cataria*), which is a member of the Lamiaceae, is attractive to members of the cat family. When the structure of the active component, nepetalactone **5.67**, was established in 1941, it was bioassayed against ten African lions from Villas Zoos in Madison, Wisconsin. The lions were reported to be aroused from a 'state of lethargy to one of intense excitement' and to becoming 'ludicrously playful' when given cotton rags soaked in a dilute solution of the lactone. There is no record of the fate of the research students. Another member of the Lamiaceae, *Coleus canina*, is alleged to deter cats from the garden.

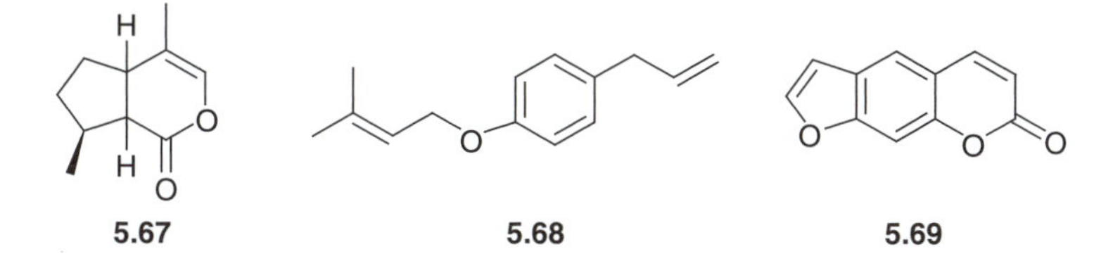

The leaves of the 'Burning Bush', *Dictamnus albus* or *D. gymnostylis*, which is sometimes known as false or white dittany, produce a substantial amount of dictagymin 1-allyl-4-(3-methylbut-2-enyloxy)benzene **5.68**. This compound decomposes easily into chavicol and isoprene

(2-methylbutadiene). The release of the low boiling (34 °C) isoprene, particularly on a hot windless day, can lead to the formation of an isoprene cloud around the plant. Isoprene has a very low flash point and the cloud can inflame without significant damage to the plant. It has been suggested that this is the 'Burning Bush' mentioned in Exodus 3 although there is some doubt as to whether this particular plant grows in Sinai. Amongst its other constituents is an alkaloid, dictamine, and the furanocoumarin psoralen **5.69**. This furanocoumarin is quite widespread in plants. Absorption by the skin and subsequent exposure to sunlight can cause a severe dermatitis. This type of biological activity is discussed in the next chapter.

CHAPTER 6

Bioactive Compounds from Ornamental Plants

Today we grow ornamental garden plants for the aesthetic pleasure that they give. This was not always so. In the Middle Ages and later, the garden provided not only pleasure but also a source of food and beneficial herbal remedies as well as some insect deterrents and even some poisons for vermin. Many of the plants which we admire for decorative purposes were introduced for practical reasons. Mediterranean herbs, particularly from the Lamiaceae (Labiatae), found their way here in the company of Roman soldiers and were also grown in the monastic gardens of the Middle Ages. The herbalists such as Nicholas Culpepper (1616–1654) record growing many plants in London which are associated with warmer countries. The Chelsea Physic Garden, which was founded in 1673 by the Society of Apothecaries, is one of Europe's oldest collections of medicinal plants and holds over 5000 species.

6.1 COMPOUNDS FROM THE LAMIACEAE

Amongst the Lamiaceae (Labiatae) there are a number of common garden plants with medicinal uses. The salvias draw their name from the Latin *salvus*, healing. There is an old saying which refers to sage (*Salvia officinalis*) and reveals the seasonal production of plant products:

He that would live for aye
Must eat Sage in May.

The monoterpenoid constituents of sage, which contribute to its use as a flavouring and in the preservation of food, have been discussed in the previous chapter. However the composition of the essential oil from the leaves differs between the varieties and with the seasons. The 'Dalmation sage' contains the monoterpenes thujone **6.1** (35–60%), 1,8-cineole **6.2** (15%), camphor **6.3** (15%), borneol **6.4** (15%) and various bornyl esters together with α-pinene **6.5**. The subspecies *S. officinalis* ssp. *lavendulifolia* contains very little thujone but rather more 1,8-cineole (29%) and camphor (34%). Greek sage, *S. triloba*, contains predominantly 1, 8-cineole (64%) together with much smaller amounts of camphor and thujone. Thujone is reported to be neurotoxic in large doses and to cause convulsions. It was the component of absinthe which caused major problems in France in the early part of the twentieth century. Sage also produces a series of phenolic diterpenoids such as carnosol **6.6** which are anti-oxidants. Sage tea is reputedly good for the treatment of gout.

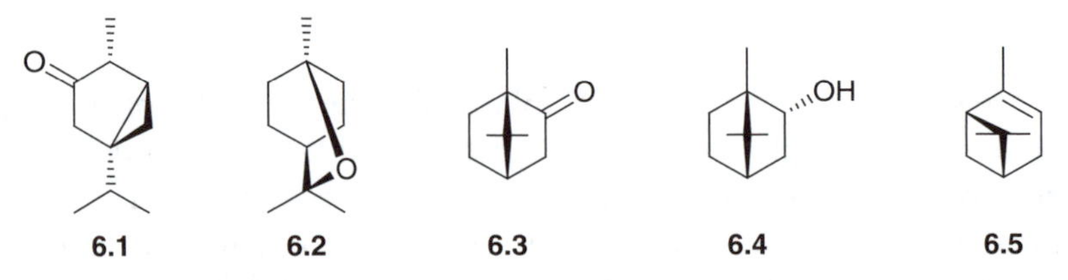

The use of the Clary Sage (*S. sclarea*) is described in various herbals such as that of Gerard (1597). It has an ambergris odour and it is grown commercially for the production of the diterpene sclareol **6.7**. This is used for the preparation of sclareolide **6.8**, which is a fixative in perfumes. Some of the other constituents may contribute to another property. In the eighteenth and nineteenth centuries it was reportedly added to beer to enhance its bitterness.

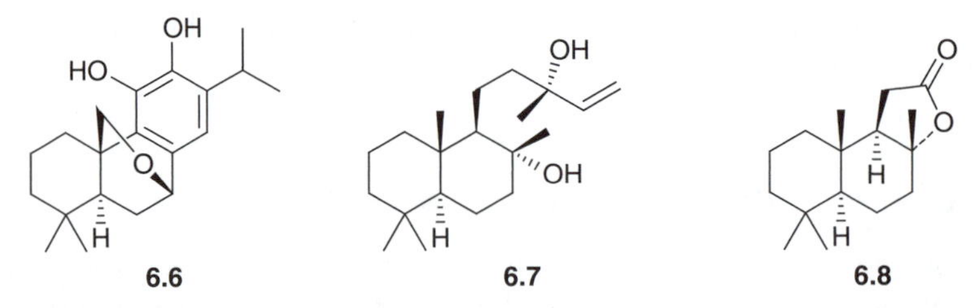

A report in 1822 states that it was 'fit to please drunkards who became either dead drunk, foolish drunk or mad drunk.' Bearing in mind the isolation of thujone from some sages and psychotropic clerodanes from other *Salvia* sp., one wonders which salvia was actually used.

Many of the salvias produce clerodane diterpenoids. Divinorin **6.9** from the Mexican species, *S. divinorum*, is responsible for the hallucinogenic properties of this plant and acts on the κ and μ opioid receptors in the brain. Divinorin and its analogues are attracting interest because they are structurally very different from the classical opioid receptor agonists of the morphine series. Some clerodanes are insect anti-feedants and in deciding which leaves to use for hallucinogenic purposes, the local Mazatecs use only those leaves that are undamaged by insect attack, *i.e.* those leaves which contain an insect anti-feedant. The common red salvia, *S. splendens*, produces some similar clerodanes such as splendidin **6.10** but there are no reports of this plant having hallucinogenic properties. The effect of this seasonal production of these anti-feedants can be seen in some salvias when the leaves are relatively undamaged prior to the flowering, but later in the season damage occurs prior to the loss of the leaf and its eventual decomposition in the autumn.

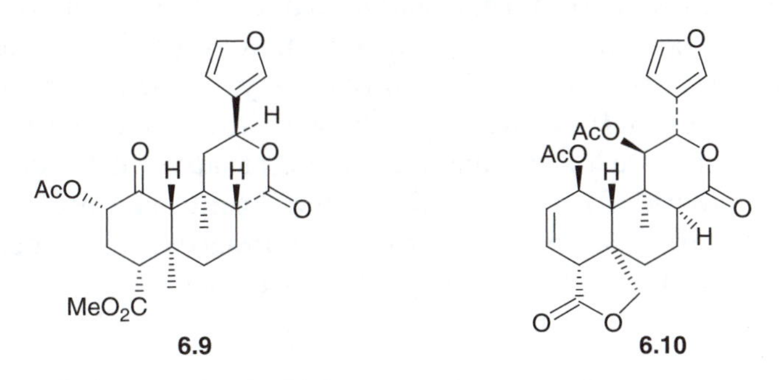

Diterpenoid quinones such as **6.11**, are produced in the roots of salvias. These compounds have anti-bacterial activity. A garden salvia with a blue flower, *S. horminum*, produces the quinone horminone **6.11** and some related phenols in the roots. The roots of a Chinese species, *S. milltiorrhiza*, are used in the Far East as the drug Tan-shen, and contain a group of diterpenoid quinones, known as the tanshinones.

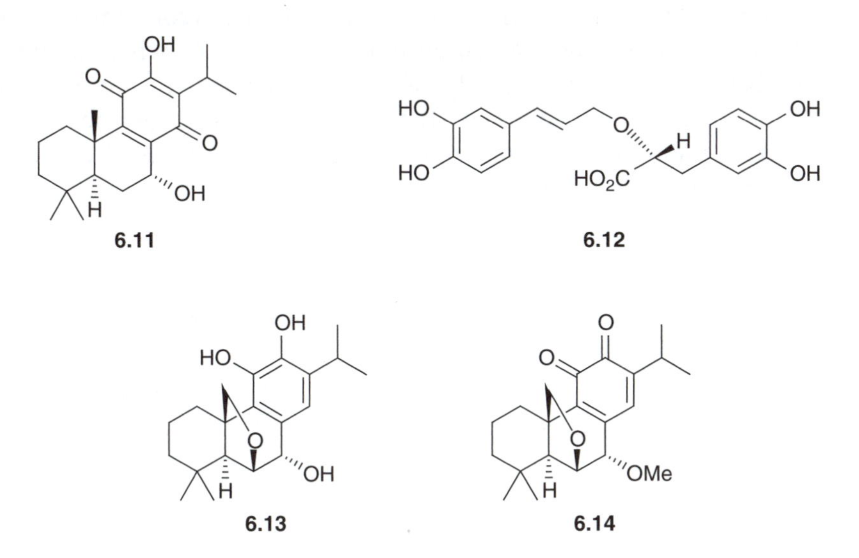

Another herb from this family that is common in the garden is rosemary (*Rosmarinus officinalis*). Although its use today is for decorative and flavouring purposes, in the past there has been considerable medicinal use and there is symbolism attached to the Shakespearean phrase 'There's rosemary that's for remembrance.' The monoterpenoids that are found in rosemary include 1,8-cineole **6.2** and α-pinene **6.4** which inhibit acetylcholinesterase and there is evidence that these monoterpenes have a beneficial effect on memory. A deficiency of acetylcholine in the brain is associated with Alzheimer's disease. The plant has stimulant and anti-oxidant properties which are associated with the presence of rosmarinic acid **6.12** and the diterpenoids carnosol **6.6** and rosmanol **6.13**. There are also some diterpenoid quinones, *e.g.* **6.14**, which have been isolated from rosemary. These diterpenoids have anti-bacterial properties and, together with the monoterpenoids, may account for the use of rosemary as a preservative in foodstuffs. The acaricidal activity of rosemary oil has been mentioned previously. The mint family are also members of the Lamiaceae and their monoterpenoid constituents have been described earlier.

6.2 THE FOXGLOVES AND CARDIAC GLYCOSIDES

Amongst the ornamental plants that have considerable medicinal use are the purple and the woolly foxgloves *Digitalis purpurea* and *D. lanata.* Although the introduction of an extract of *D. lanata* (Digitalis) into medicine for the treatment of heart failure is attributed to Wuthering in 1785, herbals indicate an earlier usage of this plant, for example, as an anti-epileptic. Digitalis is still in use today for the treatment of heart conditions. The major active compound is digoxin **6.15** which is one of the number of related cardiac glycosides. It is present to the extent of 0.1–0.4% in the plant. Whereas the steroid aglycone of the cardiac glycosides is a modified C_{21} pregnane, the leaves and seeds also contain C_{27} saponins including digitonin.

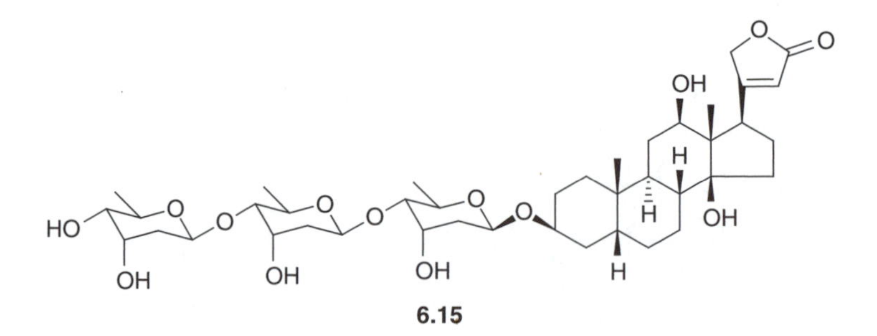

6.15

The foxglove is not the only garden plant to contain cardiac glycosides. They are also the poisonous principles found in lily-of-the-valley (*Convallaria majalis*) and the Christmas rose (*Helleborus niger*). The toxic cardiac glycosides including convallatoxin and convallarin, which are glycosides of strophanthidin **6.16**, are present throughout the lily-of-the-valley plant, including the underground rhizomes, and provide protection for the plant. Extracts have been used in folk medicine. This plant is quite aggressive, spreading by means of rhizomes. These also produce azetidine-2-carboxylic acid **6.17** which diffuses out into the adjacent soil and facilitates the dominance of this plant. Other plants absorb this unusual amino acid and mistake it for proline **6.18**. However the resultant proteins cannot function correctly and the plant dies, allowing the rhizomes of the lily-of-the-valley which can tolerate this amino acid, to spread.

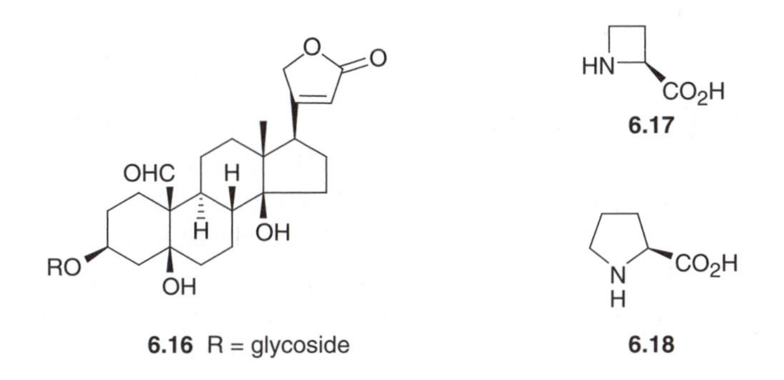

6.16 R = glycoside

6.17

6.18

6.3 POPPIES

The oriental poppy, *Papaver somniferum*, is notorious as the source of addictive narcotic, morphine **6.19** which is obtained from the resinous exudate of the calyx. Morphine is still widely used as an analgesic. The methyl ether of the phenol is codeine and heroin is diacetylmorphine. The common wild and garden poppies, such as *P. rhoeas*, are not reputed to contain much morphine although the presence of thebaine **6.20**, a precursor of morphine, has been reported. The common poppy (Flanders poppy) contains the rhoeadine alkaloids, *e.g.* **6.21**, which, like morphine, are also biosynthesized from benzylisoquinoline precursors. Greater celandine (*Chelidonium majus*) is also a member of the Papaveracae and produces some biosynthetically related alkaloids such as chelidonine **6.22** and its relative sanguinarine. These alkaloids have gastrointestinal effects and chelidonine is a mitotic poison probably accounting for the use of celandine in folk medicine to cure warts.

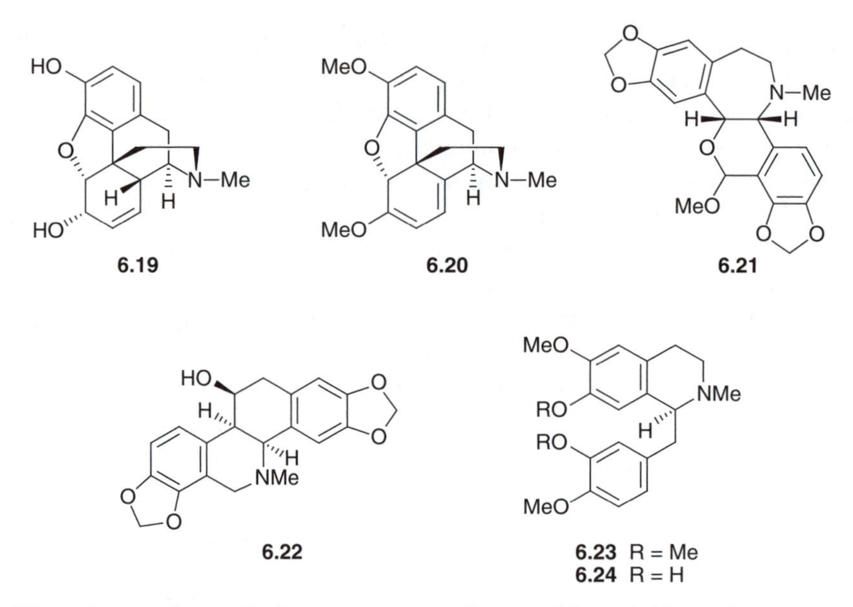

The elucidation of the structure of morphine **6.19** took over one hundred years of detailed chemistry after its first isolation in the early nineteenth century. The structure was proposed in 1923. Biogenetic speculation by Robinson in 1925 on the relationship between two families of alkaloids, the benzylisoquinoline such as laudanosine **6.23** and the morphine series, which occur in these plants, has been tested in many biosynthetic experiments in which reticuline **6.24** has played a key role. These are particularly associated with the work of Barton and Battersby in the 1960s. The key enzymatic reaction relating these compounds was identified by Barton as a phenol coupling reaction. The way in which these experiments were carried out has been described in Chapter 2.

6.4 COMPOUNDS FROM THE ASTERACEAE

Many garden flowers such as the daisies, the sunflowers, the asters and the dahlias are members of the Asteraceae (Compositae). There are several characteristic groups of natural products that have been found in these plants. The polyacetylenes are one family of compounds that has been found in these daisies and in the tubers of the dahlias. These highly unsaturated and relatively unstable compounds are exemplified by the C_{10} compounds, such as *cis*-matricaria methyl ester **6.25**. They have a characteristic ultraviolet spectrum which has facilitated their detection in plant extracts even in the presence of other ultraviolet absorbing compounds such as chlorophyll and the carotenoids. The polyacetylenes provide the plant with some protection against insect attack. Some people suffer from dermatitis on coming into contact with compounds such as falcarinol (1,9-heptadecadiene-4, 6-diyne-3-ol) **6.26**. This polyacetylene is produced by a number of plants under stress.

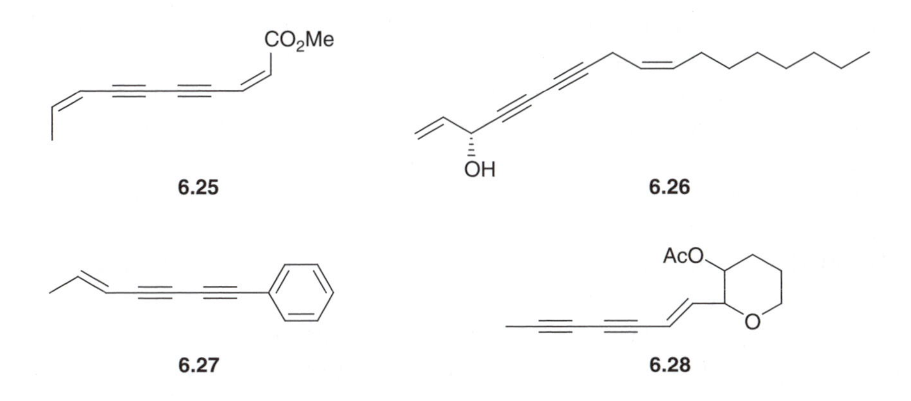

The genus *Dahlia* originated in Mexico. The many horticultural cultivars, of which about 15 000 are known, have been largely derived from two species, *Dahlia pinnata* and *D. coccinea*. Nevertheless there is quite a wide variation in the polyacetylenes that have been detected in the different cultivars. Two quite abundant examples are 1-phenylhepta-1,3-diyne-5-ene **6.27** and the pyran **6.28**. These compounds occur in the leaves, flowers and tubers in amounts as high as 2 g/kg dry weight. Not only do these compounds give protection against insect attack but some are also herbicidal and may account for the death of the grass when dahlias are grown too close to the lawn.

The African marigold (*Tagetes erecta*) has attracted interest for several reasons. The yellow-orange petals are a rich source of the carotenoid lutein which is present as a series of esters. There are reports of this carotenoid being obtained from marigolds for feeding to broiler chickens to 'improve' the yellow colour of the yolk of their eggs. The essential oil of the leaves which contains limonene, terpinolene and piperitone as the major constituents has shown anti-fungal activity against *Pythium* species associated with the 'damping-off' of young seedlings. Another valuable activity is the insecticidal action of marigolds associated with the formation of phenylheptatriyne and a series of thiophenes arising from the addition of sulfur to the alkynes. These have a useful nematocidal effect. For example there are reports of root lesions arising from nematode attack on potatoes being lower if the potato crop is preceded by marigolds. The marigold has also been recommended as an 'intercropping' plant.

The leaves and flowers of the common daisy, *Bellis perennis*, contain the polyacetylenes, deca-4,6-diynoic acid and its methyl ester as well as the monoterpenes β-myrcene and geranyl acetate. The common leaf alcohol *cis*-hex-3-enol is also present. The daisy had a reputation in folk medicine for the treatment of wounds, inflammation and rheumatism and was at one time known as 'Bruisewort'. Whilst the leaves of the daisy grow close to the ground and physically prevent the growth of competitors, there is also a possible allelopathic effect from the polyacetylenic components of the plant. The roots of the daisy contain triterpene saponins known as bellisosides. These have a hydroxylated oleanolic acid, polygalaic acid, as the aglycone. Three members of this series possess cytotoxic activity against the human leukaemia cell line HL-60 that is of the same magnitude as *cis*-platin. Dandelions (*Taraxecum* sp.) are also garden weeds whose leaves and roots have been used in folk medicine. The best known constituents of the roots are taraxacin, which is a glycoside of a taraxastane pentacyclic triterpene and the polysaccharide inulin.

Sesquiterpenoid lactones such as parthenolide **6.29** are also quite common amongst the Asteraceae. Thus parthenolide is found in tansy (*Tanacetum vulgare*) and in feverfew, *T. parthenium*, which are more often encountered as weeds. Feverfew has been recommended for the treatment of migraine headaches. The α-methylene lactones are the seat of their biological activity as cytotoxic agents. Some of these compounds can also produce contact dermatitis. *Seriphidium* (*Artemisia*) *maritimum* (sea wormwood) is used as a shrub in gardens near the sea. It produces the lactone santonin **6.30** which is responsible for the bitter taste of the plant. Santonin used to be extracted commercially for medicinal use. A number of these compounds were known as the bitter principles. The elucidation of the structure of santonin was one of the classical examples of natural product chemistry. The dienone moiety in its structure was responsible for several rearrangements under both acid-catalysed and photochemical conditions. Although it is rather toxic, as the common name of the plant suggests, it was used as a vermifuge. Another sesquiterpenoid lactone, artemisinin **6.31**, which is produced by *Artemisia annua*, is attracting interest because of its antimalarial activity. This compound was originally discovered in 1983 in a Chinese drug, Qinghaosu, which is derived from this plant. A more stable derivative, artemether (Paluther®), has been marketed since 1992.

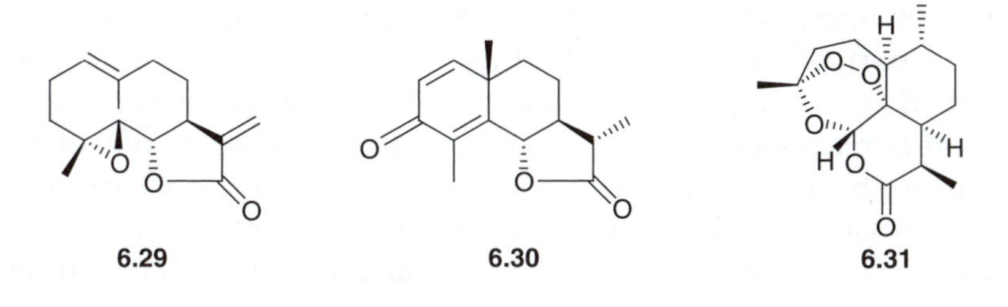

6.5 THE CONSTITUENTS OF BULBS

The constituents of various bulbs have attracted interest, particularly the alkaloids of the Liliaceae and Amaryllidaceae. The daffodil (*Narcissus pseudonarcissus*) bulb produces alkaloids exemplified by lycorine **6.32** and galanthamine **6.33**. The biosynthesis of these alkaloids involves a phenol coupling step and has been thoroughly studied. An interesting point that was exploited by the two big research groups, those of Barton and Battersby, working on this topic is that the lycorine group of natural products were produced by one daffodil cultivar ('Twink') whilst the galanthamine series were produced by another variety ('King Alfred'). A variety 'Carlton', grown under stressed conditions, has been used for the commercial production of galanthamine. Galanthamine **6.33** has attracted interest recently because it is an inhibitor of acetylcholine esterase. It is used as Reminyl® in the treatment of Alzheimer's disease. One aspect of this neurodegenerative condition is that there is a low level of acetylcholine present in the brain of those suffering from the disease. By inhibiting the biodegradation of the acetylcholine, the amount remaining in the brain increases and some of the symptoms of Alzheimer's disease are alleviated.

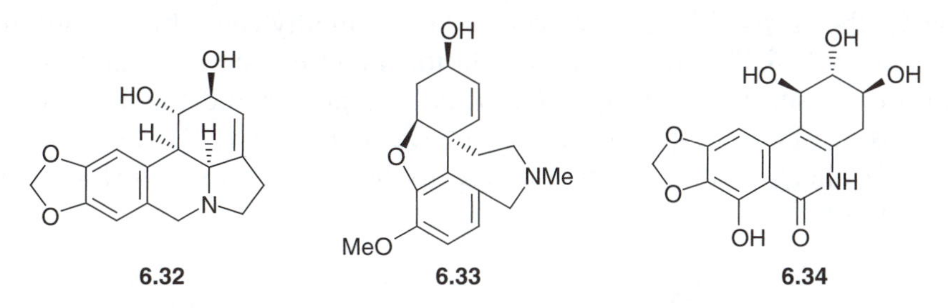

Another alkaloid produced by these plants is narciclasine **6.34**. This alkaloid has cytotoxic properties and a derivative, 7-deoxy-*trans*-dihydronarciclasine, has shown strong inhibitory activity against some cancer cell lines. When a daffodil is placed in the same vase as other cut flowers, particularly iris plants, they last longer. The effect can be replicated by placing a small amount of narciclasine in the vase. The alkaloid is responsible for delaying senescense. However this does not appear to be a general effect. Daffodil bulbs have to survive in a microbiologically hostile environment and so, not surprisingly, they produce phytoalexins in response to fungal attack. Inoculation of daffodil bulb scales with the fungus *B. cinerea* produced brown lesions in which the growth of the fungus was inhibited. Extraction of these led to the isolation of 7-hydroxyflavan, 7,4′-dihydroxyflavan and its 8-methyl analogue which possessed anti-fungal activity as phyto-alexins.

The autumn crocus (*Colchicum autumnale*) contains the alkaloid colchicine **6.35**. Although the bulbs are very toxic, extracts of the autumn crocus were known in folk medicine as a treatment for gout. The active alkaloid (0.6% dry weight) colchicine was isolated in 1884 and its unusual tropolone structure was established in 1945. Apart from its use in treating gout, colchicine has anti-cancer activity. It binds to the protein tubulin, and inhibits the formation of microtubules in mitosis.

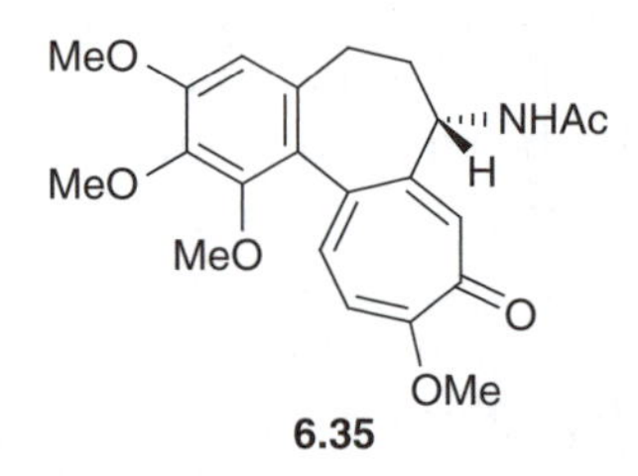

6.35

The bulbs of tulips (*Tulipa* spp.) produce a glucose ester of α-methylene-γ-hydroxybutyric acid, tuliposide A **6.36**. When the bulb is damaged, the glycoside is hydrolysed to α-methylene-γ-butyrolactone, tulipaline A **6.37**. There is a correlation between the resistance of the tulip to attack by the fungus *Fusarium oxysporum* and the presence of tuliposide. This compound has also been identified as the cause of the contact allergy, 'tulip finger' suffered by horticulturalists working with tulip bulbs.

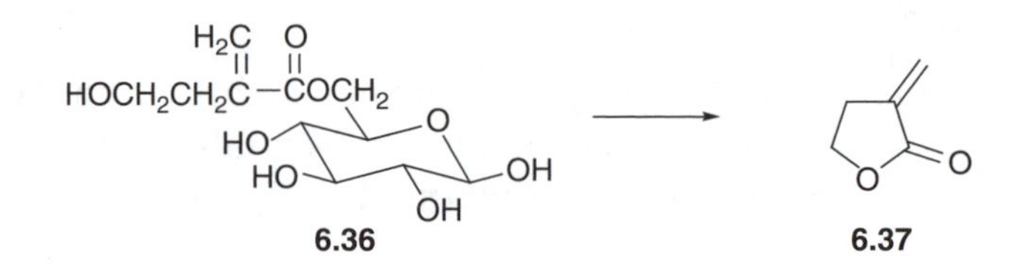

6.36 **6.37**

6.6 TOXIC COMPOUNDS FROM ORNAMENTAL PLANTS

A number of ornamental plants produce quite toxic compounds. Thus larkspur (*Consolida* (*Delphinium*) *ajacis*) and various delphiniums are noted for the presence of the poisonous diterpenoid alkaloids such as lycoctonine **6.38** which are found in the seeds. An extract of the seeds, known as stavesacre oil, was at one time used to eradicate body lice but its absorption through the skin led to serious toxicity. Similar diterpenoid alkaloids such as aconitine are produced by the aconites (monkshood).

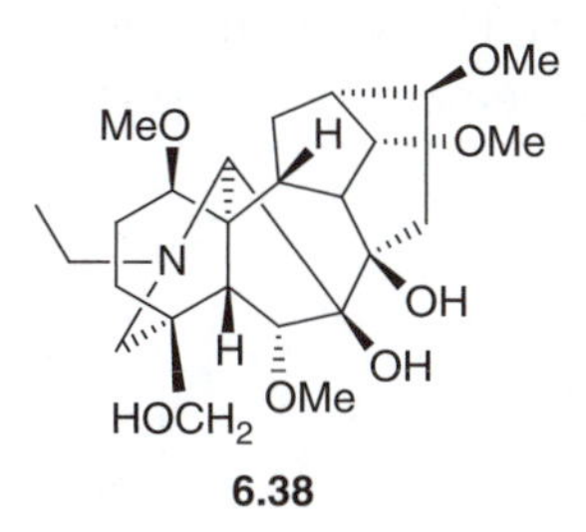

6.38

Plants from the Euphorbiaceae and Thymelaceae such as *Euphorbia globulus*, *E. characias, E. griffithi, E. amygdaloides* sp. *robbiae* and *Daphne mezereum* and the sunspurge, *E. helioscopia*, are noted for their skin irritant properties. They are quite dangerous and should be handled with gloves because the latex contains some co-carcinogenic esters. One of the worst is *Daphne mezereum*, Mezereon. It has extremely irritant constituents which are present in all parts of the plant including the berries. These compounds produce a serious reddening of the skin and, if ingested, inflammation of the mouth, throat and stomach which can be fatal. The structure of mezerein, an ester of 12-hydroxydaphnetoxin **6.39**, contains a highly hydroxylated diterpenoid skeleton with an unusual ortho ester of benzoic acid.

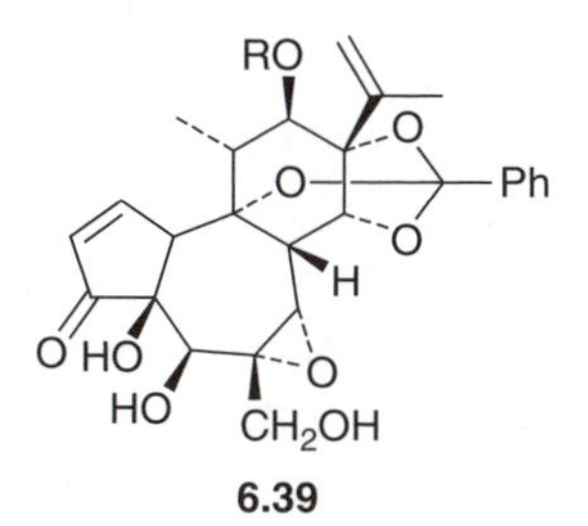

6.39

The seeds of several plants contain toxic alkaloids. The laburnum (*Laburnum anagyroides*) and lupin (*Lupinus polyphyllus*) which belong to the Fagaceae (Leguminosae) contain a series of quinolizidine alkaloids such as cytisine **6.40** and lupinine **6.41** which are responsible for a number of cases of poisoning each year. The toxicity of lupins was known to the Greeks and Romans. In the centuries before the advent of synthetic insecticides, extracts of lupins were used by the Greeks and Romans for this purpose. Many wild North American species of lupin are serious stock poisons. Their piperidine and quinazolidine alkaloid content has been associated with teratogenic effects arising from animals ingesting these alkaloids during pregnancy. The effects on the offspring of these animals include skeletal malformations (crooked calf disease) and cleft palates. The compounds inhibit acetylcholine receptors and, by acting as neuromuscular blocking agents, prevent fetal movement at important developmental stages during gestation.

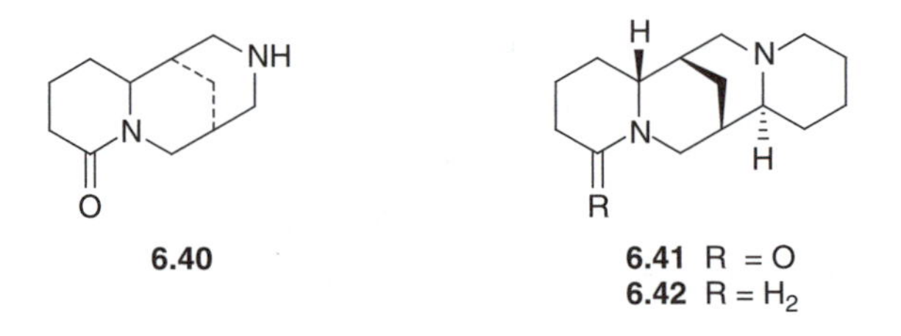

6.40

6.41 R = O
6.42 R = H_2

Those *Nicotiana* species which contain nicotine also have a useful insecticidal effect. Liquid tobacco extracts were used throughout the eighteenth and nineteenth centuries to control pests found on fruit trees and in the greenhouse. A soap solution was sometimes added to these extracts to increase their penetrating powers.

The common broom (*Cytisus scoparius* syn. *Sarothamnus scoparius*) as distinct from Dyer's Greenweed (*Genista tinctoria*) is a bush whose medicinal value was recognized in its use in heraldry. Typical of a member of the Fabaceae it produces its seeds in pods. These produce an alkaloid, sparteine **6.42** which was first isolated in 1851. Sparteine has an effect on heart rate and it also has a narcotic effect. It has been reported that sheep which have fed on the plant at first become excited and then stupefied. Broom also produces the 6,7-dimethoxycoumarin scoparin and the isoflavone genistein **6.43**. The presence of these compounds may account for the use of extracts of the plant in herbal medicines for the treatment of heart conditions. A caffeine-free herbal tea, Rooibos or Red Bush, available commercially in this country and which is made from the leaves of *Aspalathus linearis*, another member of the Fabaceae, also contains sparteine **6.42**.

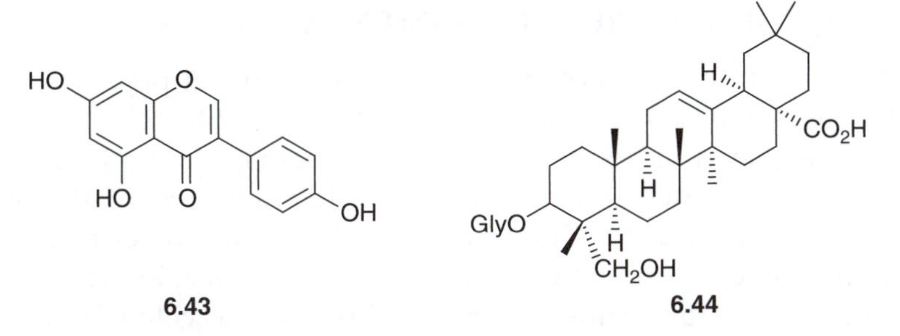

6.43 **6.44**

Another plant which can give rise to contact dermatitis is ivy (*Hedera helix*). Not only does it contain some polyacetylenes (such as falcarinol) which have this effect but also some triterpenoid saponins such as hederacoside **6.44**. The effect of these saponins, which have a combination of polar and non-polar groups in the molecule, is to act as a detergent, allowing irritant compounds to traverse the protective fat layers on the skin. Primroses (*Primula* sp.) are another group of plants which produces both saponins and, in this case, an irritant quinone, 2-methoxy-6-pentylbenzoquinone (primin) **6.45**. This can cause quite serious dermatitis in horticulturalists. Structure:activity studies have shown that the allergic reaction to primin is associated not just with the presence of the quinone but also with the length of the alkyl chain.

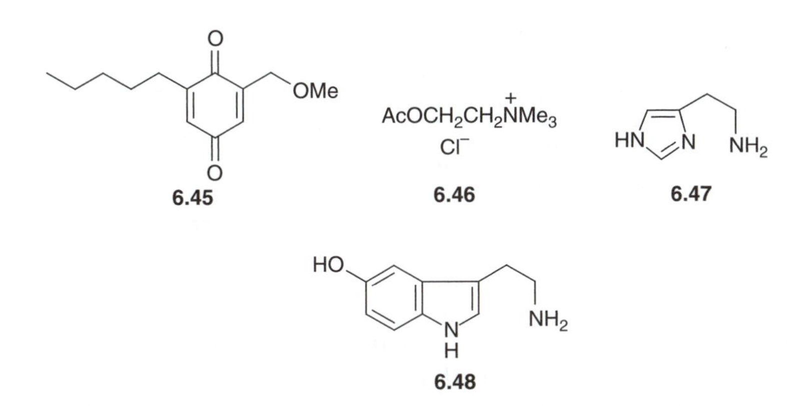

6.45 **6.46** **6.47**

6.48

The stinging nettle, *Urtica dioica*, has hairs on the leaf which possess a sharp silica tip. This punctures the skin and snaps off to release from the glandular hair small amounts of the neurotransmitters, acetylcholine **6.46**, histamine **6.47** and serotonin **6.48**. These activate the nerves to produce the stinging sensation and the red swelling characteristic of nettle stings. The presence of various astringent acids in the leaves of docks (*Rumex* sp.) may counteract the effect of these bases.

6.7 COMPOUNDS FROM ORNAMENTAL TREES

Garden trees have furnished many interesting natural products. Probably the most useful is the willow tree, *Salix* sp. The bark of the willow tree was known in folk medicine to contain a pain killer. Both the Greek physician Hippocrates and the Roman writer Pliny record the use of willow bark and leaves for the relief of fever and pain. Investigations by an Oxfordshire clergyman, the Reverend Edward Stone, in the middle of the eighteenth century, showed that dried powdered willow bark was a good remedy for fevers and led to the popular use of this treatment. The glycoside salicin **6.49** and the aglycone saligenin were isolated in 1843. Salicin and some glucoside esters such as 6′-*O*-benzoylsalicin (populin) and 2′-*O*-benzoylsalicin (tremuloidan) have been found in various poplar barks, whilst the 2′-*O*-acetate (fragilin) occurs in *Salix fragilis* (the crack willow). Salicylic acid **6.50** was introduced as a pain killer in 1876. It also has anti-microbial activity. The prolonged use of these compounds as pain killers led to the development of stomach ulcers and the less ulcerogenic compound, acetylsalicyclic acid (aspirin) **6.51**, was developed and marketed by the company Bayer in 1897. Salicylates such as methyl salicylate (oil of wintergreen) occur quite widely and are responsible for the analgesic effect of a number of plant extracts such as that of meadow sweet.

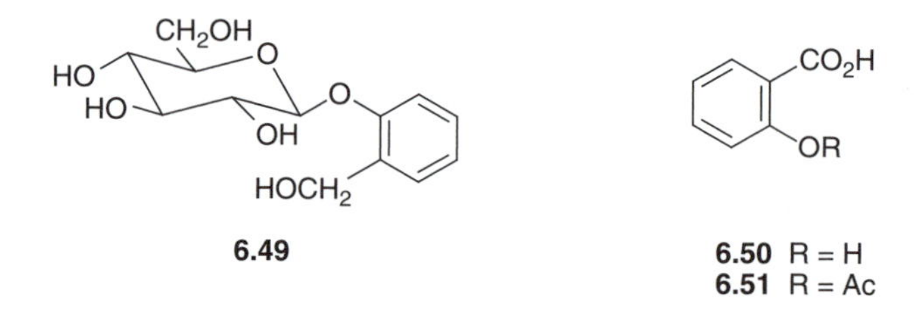

The yew tree has a chemistry which is rather different from that of other conifers and it has provided a number of alkaloids of interest. The leaves are toxic and contain a series of diterpenoid alkaloids such as taxine B **6.52**. Yew trees were often planted around churchyards to prevent grazing cattle from entering and damaging the area. In 1964 the bark of the Pacific yew, *Taxus brevifolia*, was found to contain a compound which was active against a leukaemia cell line. The structure of the diterpenoid alkaloid taxol® (paclitaxel **6.53**) which was obtained from the bark, was established in 1971. It is a very effective anti-cancer agent targeting the formation of the mitotic spindle in cell division. The use of this compound was hampered by its low occurrence and the need to destroy the slow-growing trees in order to obtain sufficient material from the bark. However taxol® can now be made by partial synthesis from more readily available taxanes such as 10-deacetylbaccatin III **6.54** which are found in the renewable parts of the plants, such as the needles of the common English yew *T. baccata*. In the search for taxanes, the phytochemistry of *Taxus* species has been thoroughly studied worldwide.

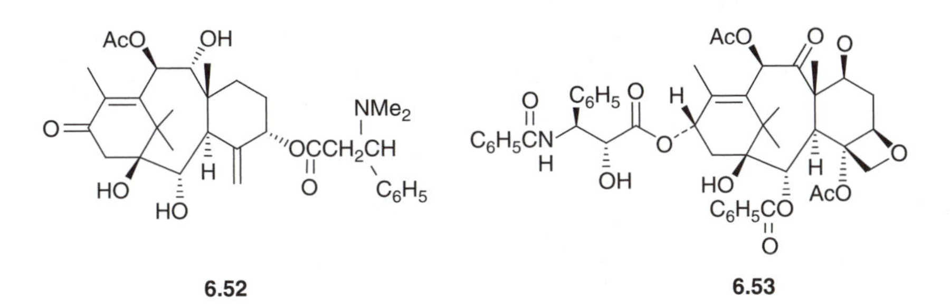

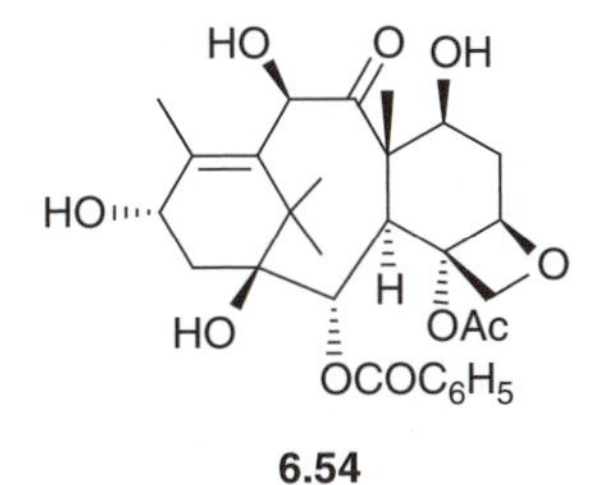

The bark and the wood of the common European pedunculate oak, *Quercus robur*, contain the polyhydroxylic phenol, ellagic acid **6.55**, which is a component of tanning agents derived from the bark. Many of the tannins which occur in the bark of trees act as feeding deterrants to phytophagous insects and grazing mammals. Some species have however become tolerant to the tannins in particular trees and then become specialized feeders on these plants. Ellagic acid is a good anti-oxidant and it inhibits the mutagenicity of benzpyrene and other aromatic hydrocarbons by blocking their epoxidation. It has anti-bacterial properties, preventing the spoilage of wine in oak casks, and it contributes to the taste of brandy that has matured in oak barrels. An oak ellagitannin, castalagin **6.56**, is eluted by wine and as it ages in the oak barrel it is converted into vescalene **6.57**. This compound has anti-topoisomerase II activity and imparts a potential anti-cancer activity to the extract.

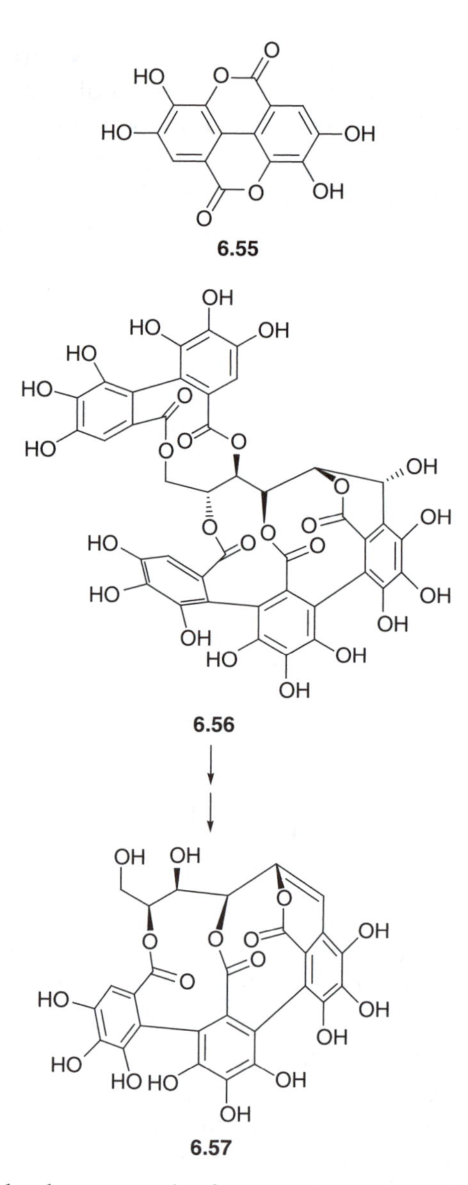

Extraction of the heartwood of many trees has yielded resins from which di- and triterpenoids and a number of quinones have been extracted. We have already come across some of the latter in the context of plant pigments. A number of the quinones are responsible for the contact dermatitis that people suffer when working with wood. Juglone **6.59** is formed in the walnut tree by hydrolysis and aerial oxidation of the 4-*O*-β-glucoside of 1,4,5-trihydroxynaphthalene **6.58**. It has allelopathic and anti-microbial activity. Very little will grow under a walnut tree.

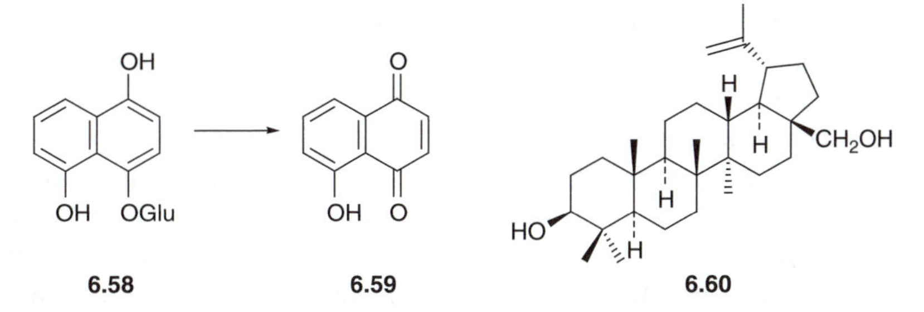

6.58 **6.59** **6.60**

The outer parts of birch bark (*Betula pendula*) contain substantial amounts (up to 20%) of the triterpene betulin (lup-20(29)-ene-3β,28-diol) **6.60**. The corresponding 28-carboxylic acid, betulinic acid, has significant activity against the HIV virus and human melanoma cells in which it inhibits phosphokinase C.

6.8 MISTLETOE

Mistletoe (*Viscum album*) is an evergreen parasitic plant which is found growing on various deciduous trees with a soft bark, typically old apple trees. Apart from the mystical properties associated with the plant, it has been used in folk medicine. However, the berries are poisonous. The plant contains the common triterpenes, oleanolic acid and betulinic acid as well as glycosides of coniferyl alcohol and sinapyl alcohols. The chalcone, 4′,6′-dimethoxy-2′-hydroxychalcone-4-*O*-glucoside and the corresponding flavanone, 5,7-dimethoxyflavanone-4′-*O*-glucoside, have also been found. The neurotransmitters, acetylcholine, histamine, γ-aminobutyric acid and tyramine together with a cyclic pentapeptide, viscumamide, comprising L-leucine units, have been reported. However much of the biological activity is associated with the presence of two groups of polypeptides, the viscatoxins with 46 amino acid residues and the larger lectins I–III. The latter may be responsible for the cytotoxic activity of mistletoe extracts. It has also been shown that mistletoe can accumulate sulfur-containing compounds such as cysteine from the xylem of its host plant.

6.9 CONIFERS

In an evolutionary context, the conifers are amongst the oldest plants and consequently their chemotaxonomy has attracted interest. Conifers are grown in gardens as evergreens and there are many cultivars of some species with different growth characteristics and shades of green. Out of the five main families, most of the garden conifers grown in the northern hemisphere belong to the Pinaceae and Cupressaceae. The Pinaceae include the genera *Abies, Cedrus, Larix, Picea, Pinus, Pseudotsuga* and

Tsuga, and include many garden trees. The Taxodiaceae include *Sequoia, Sequoiadendron* and *Metasequoia*. We have already considered the biological activity of *Taxus* species which belong to the Taxaceae. The Cupressaceae contains genera such as the *Cupressus, Chamaecyparis* and *Juniperus* which again include a number of garden trees. These botanical relationships are reflected in the chemical constituents of these trees.

The villain of many gardens is the Leyland cypress, x-*Cupressocyparis leylandii*, which often forms a rather too vigorous hedging tree that can grow as much as three feet in a year. This tree is a hybrid which was obtained in 1888 from the Monterey cypress (*Cupressus macrocarpa*) and the Alaskan cedar (*Chamaecyparis nootkatensis*) on the Leighton estate near Welshpool in Powys. The young hybrid sapling was then propagated by Mr C. J. Leyland in Northumberland, hence the name leylandii. Although the leaves contain monoterpenes typical of a *Cupressus*, the major sesquiterpene, ent-dauca-5,8-diene **6.61**, and the accompanying minor daucadienes and acoradienes **6.62** are different from those present in either of the parent species. Although the relevant cyclase gene was presumably present in one of the parents, it must have been silent and only activated in the hybrid. The leaves of x-*C. leylandii* contain some anti-fungal biflavones including amentoflavone **6.63** and hinokiflavone **6.64**. These are phenol coupling products of the flavone apigenin. Their presence accounts for some of the difficulty in composting x-*C. leylandii* cuttings. The volatile terpene content of the pines accounts for their pleasant and soothing smell.

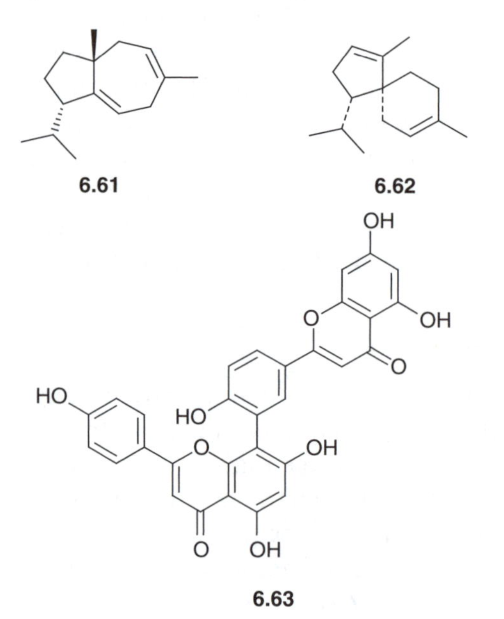

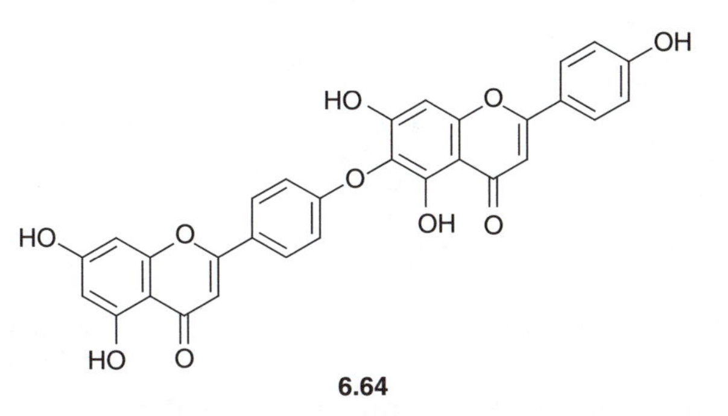

6.64

The main components of the essential oil of the Italian cypress, *Cupressus sempervirens*, are the sesquiterpenes cedrol **6.65**, α-cedrene **6.66** and 1,7-diepi-β-cedrene, and the methyl ether of carvacrol **6.67**. Cedrene is a characteristic product of *Juniperus* species and it is the major component of Virginian cedar wood oil which is obtained from *Juniperus virginiana*. Another widespread, often columnar, type of European Juniper is *J. communis*. The constituents of this conifer have been examined quite thoroughly. The monoterpenes include the phenol carvacrol **6.67** and its methyl ether and an unusal anti-fungal tropolone, α-thujaplicin **6.68**. The α-, β- and γ-thujaplicins differ from each other in the position of the isopropyl group. Apart from the monoterpenes, the sesquiterpenes cedrene **6.66** and cuparene **6.69** and the eudesmane junenol **6.70** are found in the berries. The heartwood contains some diterpenes including communic acid **6.71**, pimaric acid **6.72** and the phenols sugiol **6.73** and xanthoperol **6.74**. The latter imparts a yellowish colour to the wood.

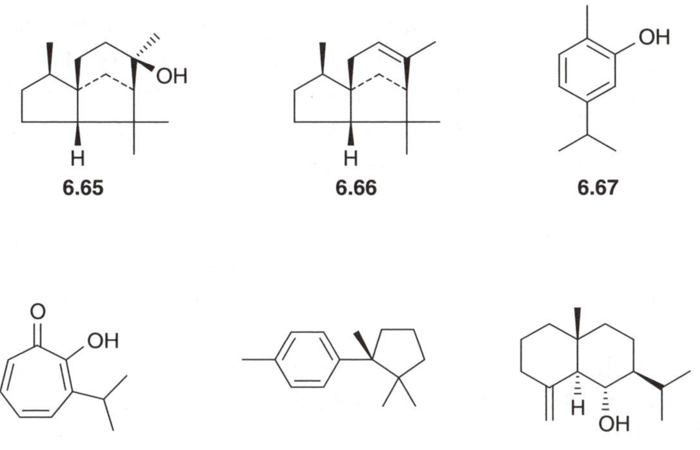

6.65 **6.66** **6.67**

6.68 **6.69**

6.70

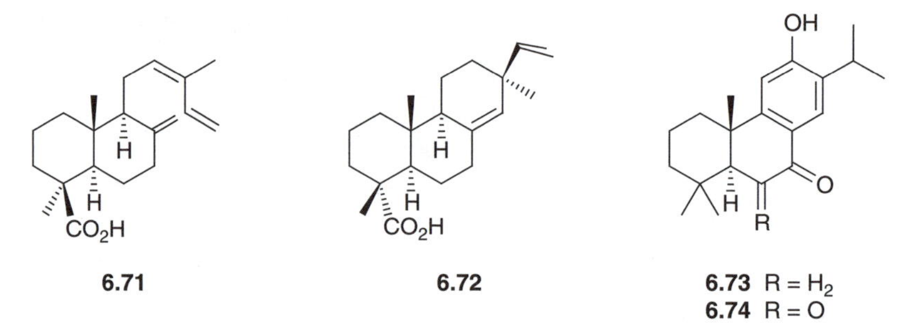

6.71 **6.72** **6.73** R = H_2
6.74 R = O

A major glucoside in many conifers is coniferin **6.75**. This is the storage and transport form of coniferyl alcohol which is used by the tree in the biosynthesis of lignin.

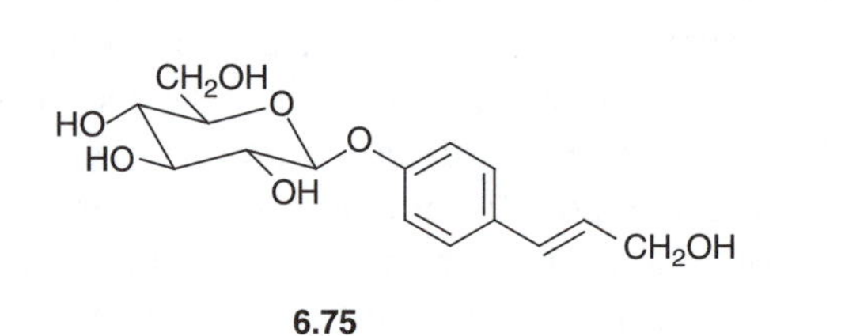

6.75

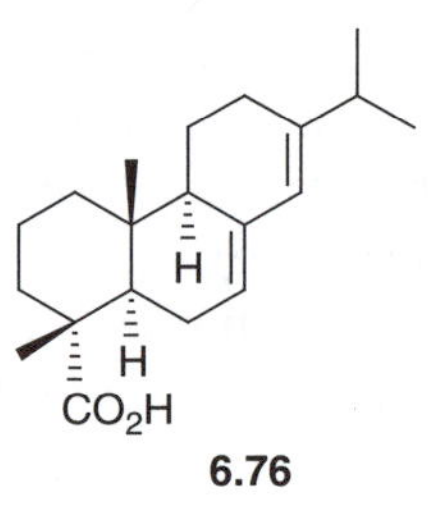

6.76

6.77 R = OH
6.78 R = H

The resin acids obtained from pines such as abietic acid **6.76** and the pimaric acids **6.72** from *Pinus* and *Abies* species play an important role in protecting the tree against bacterial and fungal decay. This property of the resin acids is used commercially in a wood resin which is supplied for protecting wood against decay. The heartwood of some species of larch have also yielded oleoresins which have found commercial use. The oleoresin from one of the larches, *Larix decidua*, which is sometimes grown as a garden tree although it is not an evergreen, contains large amounts of a diterpene larixol **6.77**. Other bi- and tricyclic diterpenes that are common include manool **6.78** and 13-epimanool.

The Christmas tree or Norwegian spruce, *Picea abies*, contains diterpenes such as abietic acid and some catechins and hydroxystilbenes such as piceatannol **6.79** and pinosylvin **6.80**. Resveratrol **6.81** has been found in the bark of *Pinus sibirica*. These compounds have quite powerful antimicrobial activity and protect the wood against decay. Some other aspects of the biological activity of the hydroxystilbenes are discussed in the next chapter.

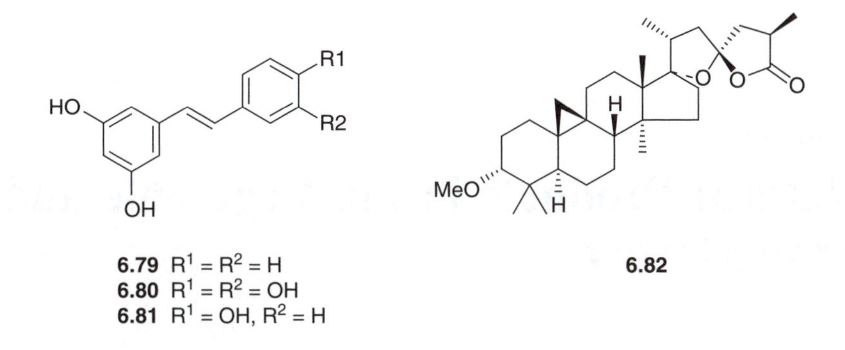

6.79 $R^1 = R^2 = H$
6.80 $R^1 = R^2 = OH$
6.81 $R^1 = OH, R^2 = H$

6.82

The oleoresins can be obtained by tapping the tree by boring into the heartwood or by scarification of the sapwood. The Scots pine, *Pinus sylvestris*, is the source of oil of turpentine. Distillation of the oleoresin gives a mixture of α- and β-pinene and lesser amounts of limonene and Δ^3-carene which is turpentine. The residue is colophony and is a mixture of abietic and pimaric acids. The bark of the silver fir (*Abies alba*) has a greyish white appearance from crystals of an abundant triterpene, abietospiran **6.82**.

Although we have described the constituents of only a small fraction of the many plants that can be grown in the garden, nevertheless there is enough to show that even a small garden has space to become a chemical herbarium of medicinal plants.

CHAPTER 7

Natural Products in the Vegetable and Fruit Garden

In this chapter we shall consider some of the compounds that contribute to the flavour and beneficial properties of common vegetables and fruits that are grown in the garden. Natural products play a major role in the nutritional value and flavour of vegetables and fruits and in their protection against pests. Various cultivars of these crops have been bred for their resistance to disease or to enhance their flavour, features which affect their natural product content. Phytoalexins may also be produced in response to microbial attack but these do not always confer beneficial properties on the plant when it is used as foodstuff. It has been suggested that some of the organoleptic differences of vegetables grown under 'organic' conditions arise from the presence of stress metabolites produced in response to microbial or insect attack.

7.1 ROOT VEGETABLES

The potato (*Solanum tuberosum*), originally a South American plant, is now one of the most widely grown root crops as a provider of carbohydrate in the form of starch. It is also a source of vitamin C, ascorbic acid **7.1** (c. 200 mg/kg).

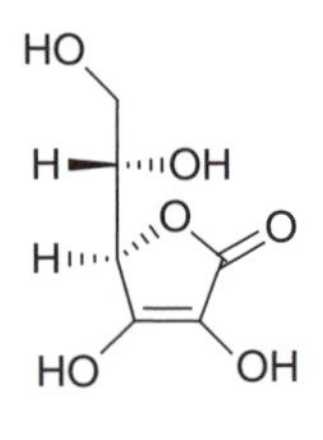

7.1

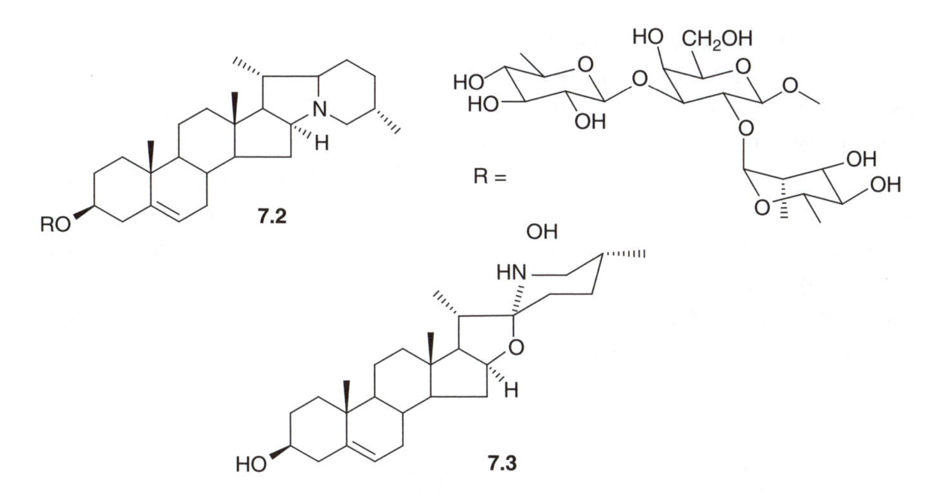

Like the tomato, the potato belongs to the Solanaceae and the potato plant contains a similar group of toxic steroidal alkaloids exemplified by solanine **7.2**. Whilst these are found mainly in the aerial parts of the plant, small amounts are found in the skin of the potato. These can increase when the skin is green and in shoots which have been exposed to light. Most of these compounds, many of which occur as glycosides, have a haemolytic effect and anti-microbial activity. Solanine also has choline esterase inhibitory activity. These alkaloids are toxic to man. It has been found that potatoes infected with the fungus *Phytophthera infestans* have a teratogenic effect in test animals. This has been associated with a change in their steroidal alkaloid content and the biological activity has been associated with the presence of α-solasodine **7.3**. In the plant the alkaloids play a role as insect feeding deterrents. However the Colorado beetle (*Leptinotarsa decemlineata*), which can be a serious pest on the potato, is not affected by solanine. A related plant, *S. demissum*, is resistant to attack and this has been associated with the presence of the steroidal alkaloid demissine. The subtle structural differences from solanine **7.2** include the presence of four sugars rather than three and the reduction of the Δ^5-double bond. Efforts have been made to breed potato plants containing this alkaloid in the leaves. Although the Colorado beetle is not yet an established pest in the UK, there has been considerable work done on its relationship with the potato in those countries where it is a pest. It is attracted to the potato by volatile sesquiterpenes, particularly caryophyllene. The aggregation pheromone which is produced by the male, is a highly oxygenated monoterpene, (*S*)-1,3-dihydroxy-3,7-dimethyl-6-octen-2-one **7.4**. Bioassays of synthetic material have revealed that the (*S*)-isomer is attractive to both male and female beetles while the enantiomer is inactive.

The widespread use of insecticides in the USA has led to resistance developing in the Colorado beetle although it is susceptible to microbial insecticides based on *Bacillus thuringiensis* (Bt). The maggots of a tachinid fly, *Myiopharus doryphorae*, are known to parasitize the larvae of the Colorado beetle and may be useful as biocontrol agents. The potato root eelworm, *Heterodera rostochiensis*, is another cause of economic loss. At nanomolar concentrations, solanoeclepin A **7.5**, which is produced by the potato, stimulates the hatching of this potato cyst nematode. The structure of this highly oxidized triterpene was established by X-ray crystallography in the 1990s.

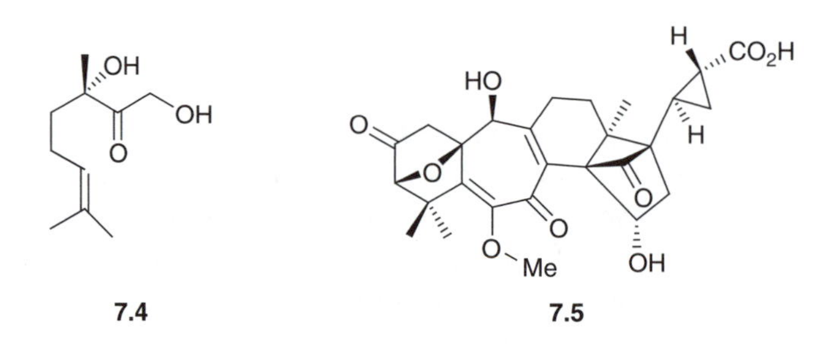

The sterols that have been identified in the leaves of the potato plant include the sequence from cycloartenol to cholesterol and are precursors of the alkaloids.

The formation of the tubers is controlled by photoperiod and it is stimulated by the production of tuberonic acid **7.6** in the leaves. This relative of the hormone jasmonic acid **7.7** is translocated to the roots from the leaves in order to stimulate tuber formation. Potato tubers do not sprout immediately after harvest but experience a period of winter dormancy before they are 'chitted' and buds appear. However treatment of freshly harvested potatoes with a solution of the plant hormone gibberellic acid shortens this rest period. In some countries it is possible to obtain two crops in a year using this treatment. The gibberellins GA_1 and GA_{20} have been detected as endogenous plant hormones in potato sprouts.

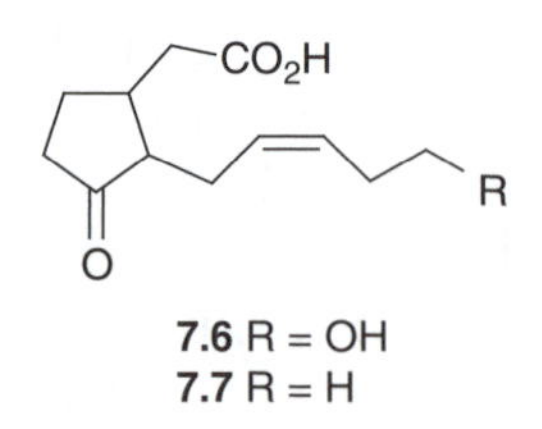

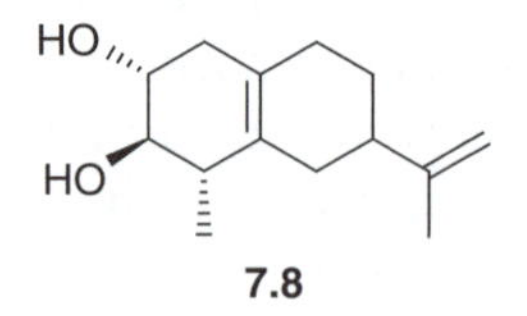

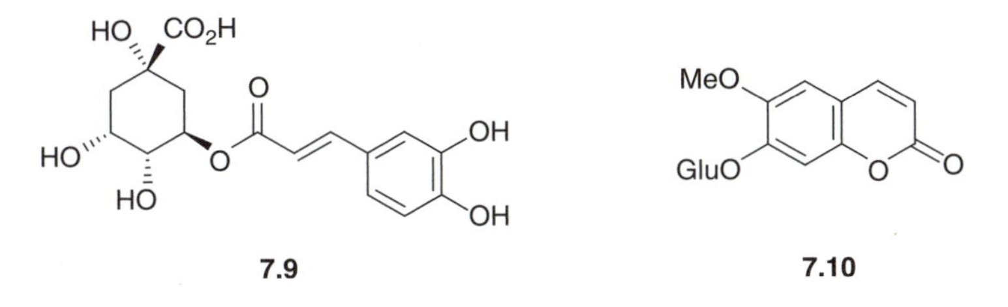

7.9 **7.10**

When the tubers are attacked by a fungus such as *Phytophthora infestans*, they produce a sesquiterpenoid phytoalexin rishitin **7.8**, which is a natural anti-fungal agent. The tubers also respond to damage by insects by producing a polyphenol oxidase enzyme system. The tubers contain a number of polyhydroxylic phenols including chlorogenic acid **7.9** which is formed from dihydroxycinnamic acid (caffeic acid) and quinic acid. The amounts of these increase around a wound. The polymeric oxidation products of these phenols, which arise from the action of the polyphenol oxidase, seal the area of the wound and produce the characteristic browning reaction. The coumarin glycoside scopolin **7.10** is also produced and is a powerful anti-fungal agent.

Traces of iron(II), absorbed by the potato, form a complex with chlorogenic acid and this can be oxidized to the iron(III) complex on boiling, giving rise to the slightly grey discolouration of some cooked potatoes. However the presence of these small amounts of polyhydroxylic phenols in the potato confer considerable beneficial effects as anti-oxidants.

The carrot (*Daucus carota*) is another popular root crop. Carrot seeds are small and there is a temptation to sow them very close together. However carrot seed has been reported to release crotonic acid into the soil which above a certain concentration, can inhibit the germination and development of seedlings. This natural means of controlling plant numbers, presumably in order not to exhaust nutrients, is also a warning not to waste seed. On the other hand it has been observed that many seeds germinate after a bush fire. It was found that smoke derived from burning plants can stimulate the germination of a wide range of seeds. This has led to the isolation and synthesis of 3-methyl-2(*H*)-furo-[2,3]-pyran-2-one **7.11** as the active principle.

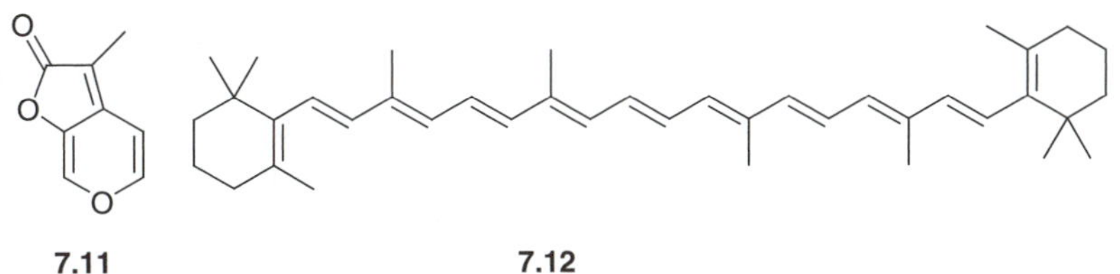

7.11 **7.12**

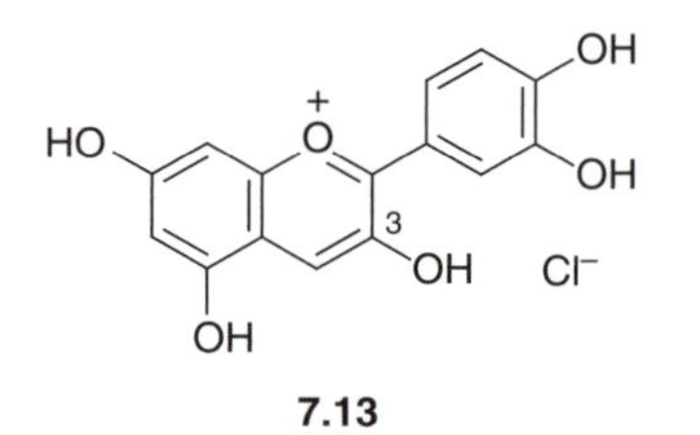

7.13

The carrot has given its name to the yellow-red carotenoid pigments. The major component of the carrot is β-carotene **7.12**. A yellow complex with a protein has been detected. The colouring matter of the purple carrot is an anthocyanin, a glycoside of cyanidin **7.13**. The carrot flavouring comprises monoterpenes including sabinene **7.14**, myrcene **7.15** and the aromatic hydrocarbon *p*-cymene. The sesquiterpenes caryophyllene **7.16** and humulene **7.17** contribute to the spicy and woody background of the flavour, whilst the sweet taste comes not just from sugar but also from β-ionone **7.18**, a carotenoid degradation product. The sesquiterpenoids carotol **7.19** and daucol **7.20** are also found in carrot seed. The isoprene units of the monoterpenoids are biosynthesized by the deoxyxylulose pathway whilst the sesquiterpenes arise from both the deoxyxylulose and mevalonate pathways.

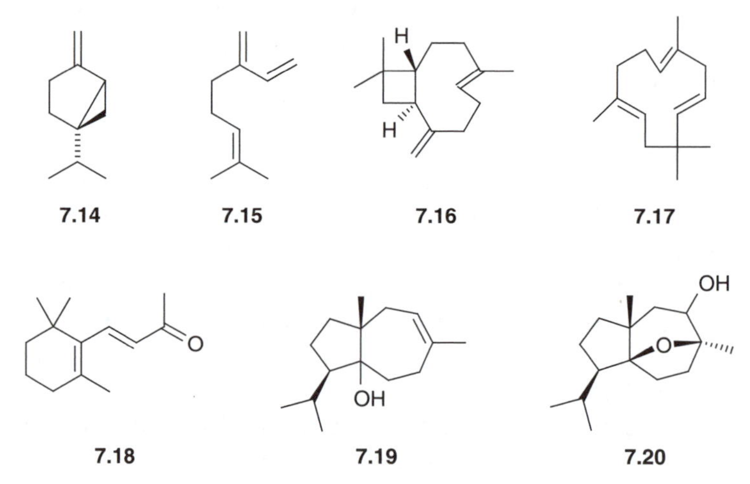

7.14 **7.15** **7.16** **7.17**

7.18 **7.19** **7.20**

Carrots produce anti-fungal polyacetylenes such as falcarindiol **7.21** in response to attack by micro-organisms. This compound and another stress-induced metabolite, 6-methoxymellein **7.23**, are responsible for the bitter taste of infected carrots. Despite the bitter taste, the polyacetylenes falcarinol **7.22** and falcarindiol **7.21** have some beneficial effects. Apart from their anti-bacterial activity, they show cytotoxic activity against some human cancer cell lines and may have a chemopreventive action against various tumours. Although carrots belong to the Apiaceae (Umbelliferae), the cultivated varieties do not produce the characteristic coumarins under normal conditions although 6-methoxymellein **7.23** is produced as a result of microbial attack.

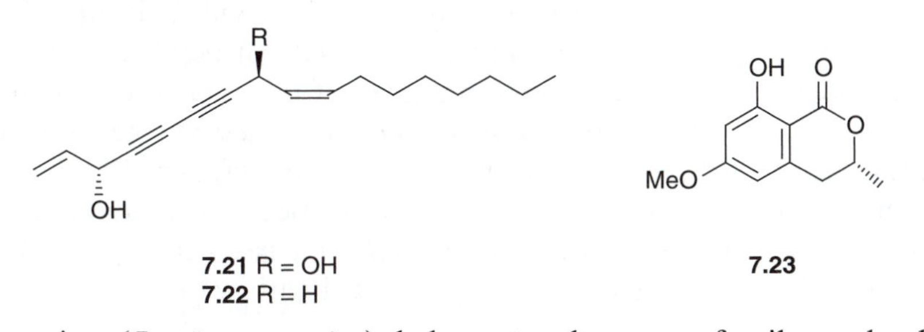

7.21 R = OH
7.22 R = H

7.23

Parsnips (*Pastinaca sativa*) belong to the same family and when infected by fungi produce the furanocoumarin psoralen **7.24**. Compounds of this type when absorbed by the skin and activated by sunlight produce contact dermatitis. There have been a number of reports of contact dermatitis arising from handling parsnips that have been infected. Similar problems arise from weeds such as cow parsley and the giant hogweed which also belong to the Apiaceae. Although it is not a root vegetable, celery is also a member of this family. When celery (*Apium graveolens*) is infected by fungi, it produces furanocoumarins which spread throughout the plant. The flavour of celery comprises four major components, 3-butylphthalide **7.25**, 3-butyltetrahydrophthalide, apiole **7.26** and myristicin **7.27**. Celery seed oil, which is used in flavourings, has a different composition and contains the monoterpene limonene **7.28** and the sesquiterpene β-selinene **7.29**, as well as 3-butylphthalide **7.25** and its dihydroderivative sedanenolide **7.30**.

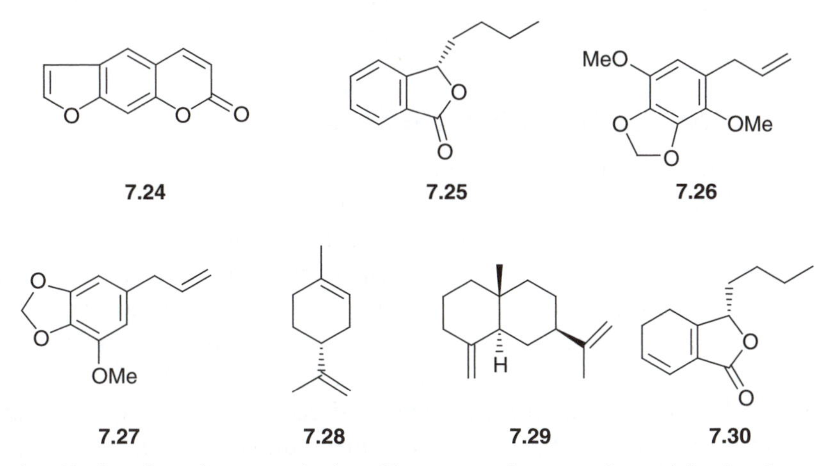

7.24 **7.25** **7.26**

7.27 **7.28** **7.29** **7.30**

The distinctive characteristic of beetroot (*Beta vulgaris*) is the deep red water-soluble pigment betanin **7.31**. This pigment belongs to the class known as the betalains which are iminium salts of betalamic acid **7.32**. Betanin is a glycoside which is also found in a number of other plants. There are albino varieties of beetroot which lack this pigment and some yellow-orange coloured varieties which contain vulgaxanthin **7.33** in place of betanin. This is the iminium conjugate of betalamic acid with glutamine rather than an indole. Pigments of this type also provide the red colouring matter of the fungus *Amanita muscaria* (the fly agaric). Another characteristic of beetroot is the 'earthy' flavour. This has been attributed to the presence of geosmin **7.34**. Beetroot contains 5–10% sucrose whilst the sugar beet variety (*B. vulgaris* var *esculenta*) may contain 15% sucrose which can be extracted with water.

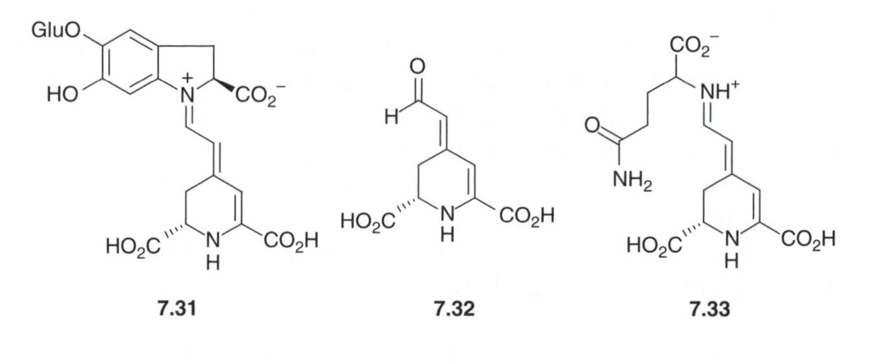

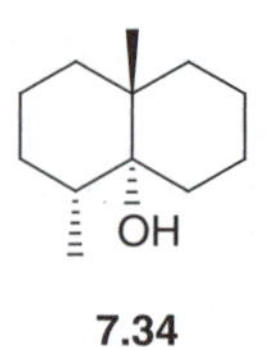

7.2 ONIONS, GARLIC AND ASPARAGUS

Onions, shallots, garlic, leeks and chives are all members of the genus *Allium*. Much has been written concerning these plants over many centuries with some authors condemning their odour and others extolling their beneficial effects. A significant part of the chemistry of these vegetables is determined by the presence of sulfur-containing amino acids and their metabolites. Several aspects of sulfur chemistry contribute to the properties of *Allium* metabolites. Sulfur is a poorer hydrogen bond acceptor than oxygen and consequently its compounds are more lipophilic and volatile. Thiols are readily oxidized to disulfides and sulfenic acids whilst alkylsulfides give sulfoxides, sulfones and thiosulfonates. Sulfur compounds also undergo some distinctive rearrangements. Finally thiols and sulfides are powerful nucleophiles and readily form metal complexes.

The chemistry of the onion, *Allium cepa*, has attracted considerable interest because of the organoleptic and beneficial properties of its constituents. Onions contain a range of organosulfur compounds that are biosynthesized from the sulfur-containing amino acid *S*-propenylcysteine *S*-oxide **7.35**. This amino acid is stored as γ-glutamyl-*S*-1-propenylcysteine. The enzyme systems such as allinase that are responsible for the production of some volatile constituents, are held in separate compartments to their substrates and only mixed when there is damage to the onion bulb. The lachrymatory component of the onion which is formed is thiopropanal-*S*-oxide **7.36**. Onions contain allylpropyldisulfide **7.37** and this has a beneficial effect in lowering blood sugar and cholesterol levels. Onions also produce the cepaenes such as **7.38** which are inhibitors of platelet aggregation in the blood. These compounds probably function by inhibiting the enzyme system cyclooxygenase which is responsible for the formation of the prostaglandins and prostacyclins. The prostacyclins are mediators of platelet aggregation. The inhibitory action of these compounds on prostaglandin biosynthesis also reduces inflammation. The dipeptide, γ-L-glutamyl-*trans*-1-propenyl-L-cysteine sulfoxide, which is present, inhibits the formation of osteoclasts and reduces bone loss.

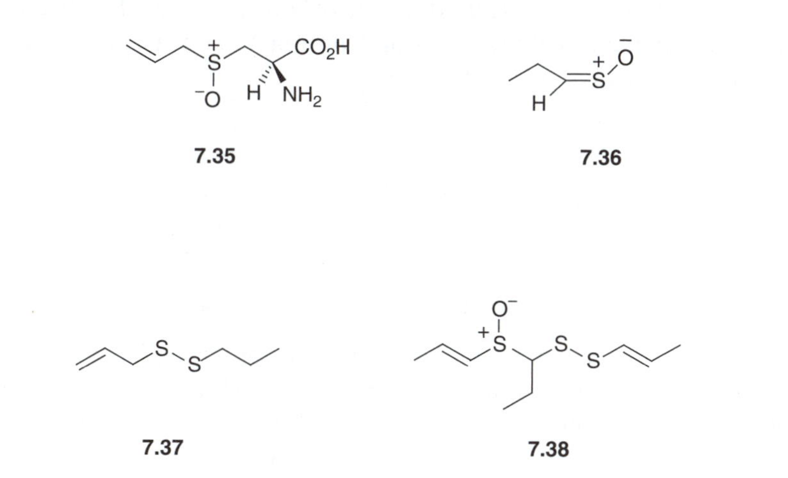

The flavonoids which have been detected in onion are quercetin **7.40**, which is the major component, kaempferol **7.39**, myrcetin **7.41** and catechin **7.42**. Their presence has been correlated with the anti-oxidant and anti-proliferative activity of onion.

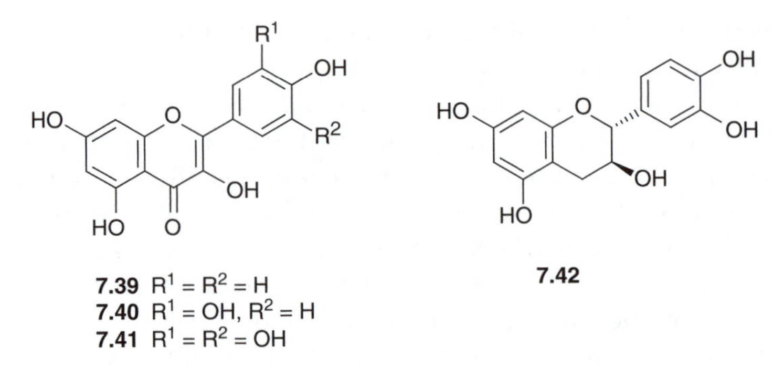

The major medicinal compound that is obtained from garlic is allicin **7.43**. It is formed when garlic cloves are damaged leading to a combination of the enzyme allinase and the amino acid alliin (*S*-propenylcysteine sulfoxide) **7.35**. Allicin has powerful anti-bacterial and anti-fungal activity. It disproportionates to give diallyldisulfide and sulfur dioxide. The former is responsible along with allylmethylsulfide, diallylsulfide and hydrogen sulfide for the garlic breath odour. Diallyltetrasulfide also inhibits the growth of the bacterium *Helicobacter pylori*, which is the causative organism of some stomach ulcers, suggesting a possible protective action of garlic. Garlic can also incorporate selenium to form selenocysteine. Organoselenium compounds such as dimethylselenide, have been detected in human breath after ingestion of this garlic. The essential oil of garlic containing a mixture of allylmethylsulfide and diallylpolysulfides, impregnated on absorbent granules, has been shown to have a repellent activity to small animals such as moles and crop damaging birds such as starlings.

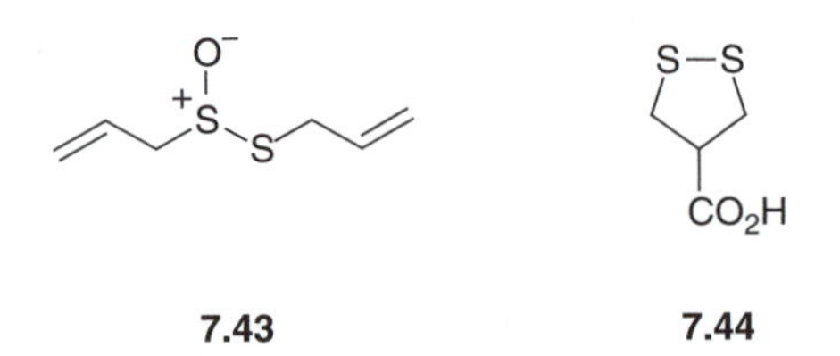

7.43 **7.44**

Asparagus (*Asparagus officinalis*) is a popular delicacy. It was the original source of the amino acid asparagine, which was first isolated in 1806. Pharmacological studies on the plant have demonstrated anti-inflammatory and cytotoxic activity. Asparagus is a member of the Liliaceae family which also includes the *Alliums*, onions and garlic. Not surprisingly a number of sulfur containing compounds derived from cysteine have been isolated from asparagus. These include *S*-(2-carboxy-*n*-propyl)-L-cysteine and some disulfides which contribute to the flavour. The cyclic disulfide asparagusic acid **7.44** which was isolated, has growth inhibitory effects on, for example, lettuce seedlings. The methyl esters of the epimeric sulfoxides of asparagusic acid, in which the *S*-oxide takes up either a *syn* and *anti* configuration relative to the methyl ester, have also been found in asparagus. The unpleasant smell in the urine after eating asparagus arises from the presence of *S*-methylprop-2-enethioate which is probably a metabolite of asparagusic acid. There are also some unusual acetylenes such as asperenyne **7.45**, some glycerol derivatives including 1,3-*O*-di-*p*-coumaroylglycerol, the sesquiterpene blumenol C **7.46** and a number of flavonoids which have been isolated. The roots of the asparagus plant have been shown to contain steroidal saponins in which sarsasapogenin and its 17α-hydroxyl derivative are the aglycones.

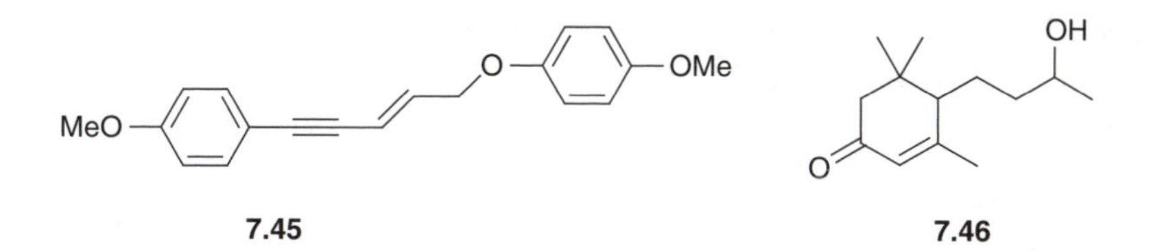

7.3 THE BRASSICAS

The aerial parts of green vegetables such as broccoli and cabbage, which are subspecies of *Brassica oleracea*, contain some interesting and useful natural products. Several of these provide protection as anti-oxidants in the control of degenerative diseases. Compared to cabbage, broccoli contains relatively high levels of β-carotene **7.11** (0.89 mg/100 g fresh weight), α-tocopherol **7.47** (1.62 mg/100 g) and vitamin C **7.1** (74.7 mg/100 g) as well as a family of glucosinolates. The tocopherols, which are present in many plant and seed oils, are important for their radical scavenging abilities.

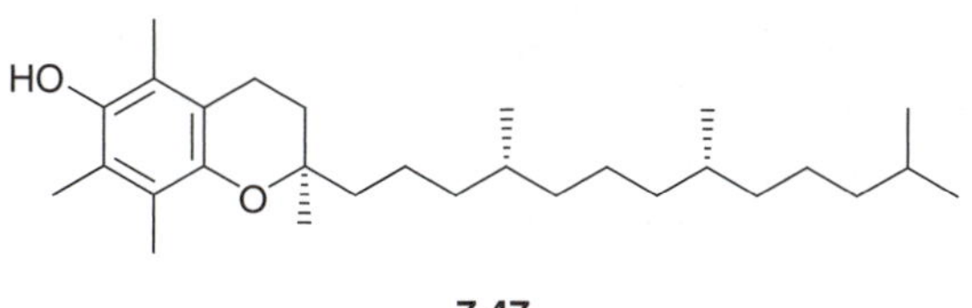

7.47

As the botanical name for broccoli (*B. oleracea*) suggests, the leaves of these plants contain quite a high oil content, typically a C_{29} hydrocarbon such as nonacosane. Oil seed rape, *B. napus*, is a relative. These oils will steam distil onto a saucepan lid when broccoli is cooked. The glycerides of a number of unsaturated fatty acids are also present.

A characteristic group of natural products which are produced by members of the Brassicaceae are the glucosinolates. These compounds have the general formula **7.48**. There are quite wide variations in the glucosinolate content of different cultivars of *B. oleracea*. Much of the biological activity of the glucosinolates arises from their hydrolysis products. Hydrolysis is catalysed by the enzyme myrosinase, which is present in compartments in plant tissue that are separate from the substrates. As with many other plants, disruption of the tissue brings the enzyme and substrate into contact and the bioactive products are formed. The isothiocyanates that are formed are sometimes known as the mustard oils. The oil which is used in the preparation of mustard is obtained by the steam distillation of the seeds of *Brassica nigra* or *B. juncea*. Radishes (*Raphanus sativus*) owe their pungent taste to the presence of glucosinolates.

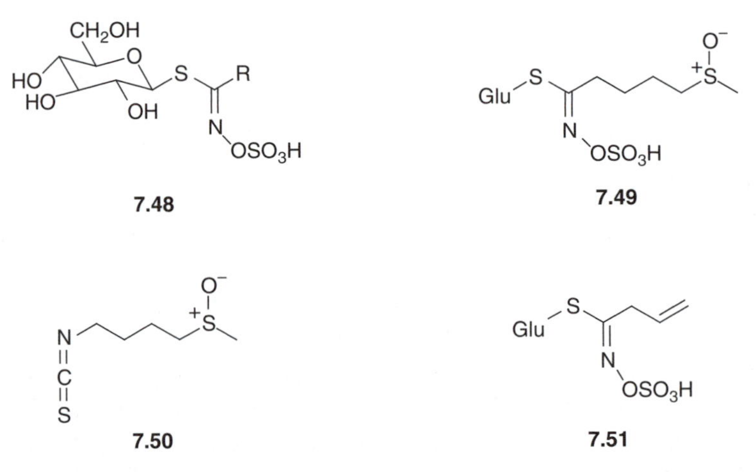

Cleavage of glucoraphanin (4-methylsulfinylbutylglucosinolate) **7.49** by myrosinase in broccoli gives rise to sulforane (4-methylsulfinylbutyl-isothiocyanate) **7.50** which has several beneficial properties. The metabolism of compounds derived from foodstuffs or medicines by the liver can be divided into two phases. In phase one the compound may be hydroxylated or epoxidized and in phase two a water-solubilizing or detoxifying molecule is attached to the metabolite. Sulforane **7.50** induces phase two enzymes in the liver which conjugate the tripeptide glutathione with potential toxic metabolites. This conjugation can detoxify electrophilic carcinogens produced by oxidative phase one metabolic changes. Sulforane also inhibits the growth of the bacterium *Helicobacter pylori*, which is associated with the formation of ulcers in the stomach. Myrosinase is partially destroyed by boiling and completely destroyed by the use of microwaves.

Sinigrin **7.51**, which is found in cabbage, gives rise to allylisothiocyanate. This isothiocyanate is a powerful anti-fungal agent acting against the fungus *Peronospora parasitica*, which causes a powdery mildew on cabbage. The volatile isothiocyanates contribute to the smell of cabbage. Sinigrin is toxic to insects and acts as a feeding deterrent to many butterflies. However the cabbage white butterfly, *Pieris brassicae*, has learnt to tolerate sinigrin and uses it as an oviposition stimulant. This provides the cabbage white caterpillar with an undisputed source of food. The relationship between the cabbage white butterfly and its host is discussed in the next chapter. Glucobrassicin is a glucosinolate which contains a 3-indolylmethyl residue. This and indolylacetic acid may be the source of indole-3-carboxaldehyde which has been isolated from cabbage.

An unwanted glucosinolate that is present in a number of the brassicas including cabbage, is progoitrin **7.52**. Cleavage of this by myrosinase produces the toxic oxazolidinethione, goitrin **7.53**. White cabbage can contain 20 mg/kg in the autumn. Goitrin reduces the ability of the thyroid gland to produce sufficient triiodothyronine and thyroxine. The gland may enlarge to compensate for this, leading to 'cabbage goitre'. 'Eat up your cabbage' may not be the wisest command to children.

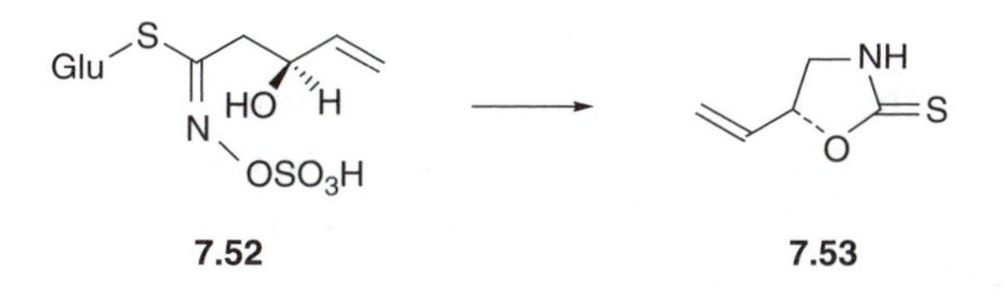

7.4 LETTUCE

Lettuce (*Lactuca sativa*) is a member of the Asteraceae. As a green salad vegetable, it is considered to be a healthy food. A part of this arises from the anti-oxidant flavonoid content exemplified by quercetin 3-*O*-galactoside and glucoside. Cyanidin 3-*O*-glucoside and its 3-*O*-(6-*O*-malonylglucoside) are found in the red-leafed variety. The dicaffeoyl ester of tartaric acid, chicoric acid, has been found both in lettuce and in chicory (*Cichorium intybus*). The carotenoids, β-carotene and lutein, also contribute to the anti-oxidant activity. Other terpenoids that have been found in lettuce include the triterpene lupeol and the sterol β-sitosterol.

Typical of the Asteraceae, lettuce produces several sesquiterpenoid lactones. The major lactones that are found in lettuce are lactucin **7.54** and lactucopicrin. They have been shown to occur as their 15-oxalate and 8-sulfate esters. These lactones are bitter tasting and in many lettuce cultivars, their contribution in the leaves has been reduced. However they are found in the milky latex which is exuded from damaged stems and roots. They have been shown to be powerful insect anti-feedants. Lactucin has been shown to elicit an interesting behavioural response by the larvae of the cabbage looper *Trichoplusia ni*. The larvae are induced to transect a leaf with a narrow trench before eating the distal section of the leaf, thus reducing their exposure to the latex exudate during feeding. These sesquiterpenoid lactones are present to a greater extent in chicory and, along with the plant phenolics such as cichoriin, protect the plant. The roots of chicory when dried and roasted can be used as a coffee substitute because of the bitter flavour which these compounds impart to the drink. A relative of lactucin, lettucenin A, is produced by the lettuce under the stimulus of microbial attack and is a powerful phytoalexin.

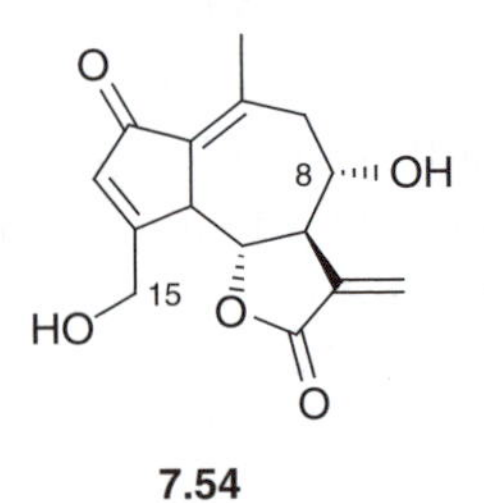

7.54

An extract of the lettuce plant known as lactaucarium was at one time used as a sedative and its activity has been associated with the presence of these lactones. Some wild *Lactuca* species, *e.g. L. serriola* and *L. virosa*, contain higher concentrations of these lactones and have been used in herbal medicine. The dried milky latex from these plants was at one time known as lettuce opium.

Rocket lettuce (arugula), *Eruca sativa*, belongs to a different family and is a member of the Brassicaceae like the cabbage. The hot taste and characteristic odour of this salad vegetable are due to its glucosinolate content. The major glucosinolate is 4-mercaptobutylglucosinolate ($R = CH_2CH_2CH_2CH_2SH$). Its dimer, S-methyl and 4-(β-D-glucopyranosyldithio) derivatives as well as sinigrin and glucoerucin have also been detected. On hydrolysis by myrosinase an isothiocyanate, 4-mercaptobutylisothiocyanate and its disulfide dimer, are formed from the 4-mercaptobutylglucosinolate and also contribute to the characteristic odour.

Water cress (*Rorippa nasturtium - aquaticum*) is also a member of the Brassicaceae. It contains the hot tasting glucosinolate, gluconasturtiin (phenethylglucosinolate, **7.48**, $R = CH_2CH_2C_6H_5$) which on hydrolysis by myrosinase gives phenethylisothiocyanate.

7.5 THE LEGUMES

Peas (*Pisum sativum*) are members of the Fabaceae. Isoflavones and their relatives are characteristic metabolites of this family of plants. The structure of the pterocarpan, pisatin **7.55**, was one of the first phytoalexins to be established in 1962. It was obtained from the endocarp of peas that had been infected with a fungus, *Monilinia fruticola*. Pisatin showed anti-fungal activity at a concentration of 10^{-4} M. The elucidation of the structure was one of the early applications of NMR spectroscopy of natural products when it was used to locate the position of the methylenedioxy group. The pterocarpan structure of pisatin has been shown to be derived biosynthetically from an isoflavone via sopherol.

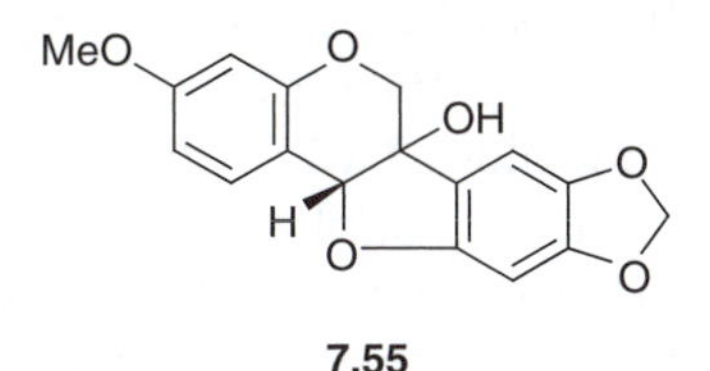

7.55

Non-proteogenic amino acids are found in some plants where they play a role in the defence mechanism. Homoserine together with two alanine derivatives, willardine **7.56** and isowillardine **7.57**, have been detected in growing pea shoots. The related non-protein amino acid β-(3-isoxazolin-5-on-2-yl)alanine, has been found in the root exudate of peas. It has growth inhibitory activity towards the seedlings of other plants such as lettuces. This allelopathic role, whilst facilitating the development of peas, may limit other crops that can be grown alongside peas. 2-Alkyl-3-methoxypyrazine **7.58** together with 2,6-nonadienal and 2,4-decadienal are responsible for the odour of green peas.

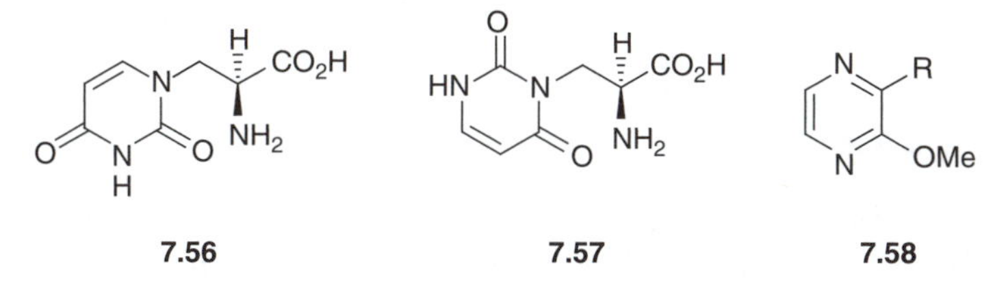

7.56 **7.57** **7.58**

Flatulence is one of the less pleasant aspects of eating peas and beans. The galactose derivatives of sucrose, known as raffinose and stachyose, accumulate in the seeds during maturation (1.2 and 3.2 g per 100 g respectively compared to 6.2 g sucrose per 100 g). The human intestinal mucosa lacks the enzyme system α-galactosidase. Consequently these oligosaccharides are not cleaved until they reach the lower intestinal tract where they are subject to bacterial metabolism by *Clostridium perfringens* and *Escherichia coli* releasing carbon dioxide, hydrogen and methane.

7.6 RHUBARB

Rhubarb (*Rheum rhaponticum*) is a vegetable that is grown for its leaf stalk (petiole) which is used as a substitute for fruit particularly in the spring. Its medicinal properties were much prized in the eighteenth and nineteenth centuries. Rhubarb is quite acidic, containing malic **7.59**, citric **7.60** and oxalic acids. The concentration of the latter is higher in the leaves (c. 0.6%) leading to the claim that these are toxic. However there are older reports that the leaves were once used as a substitute for spinach. Presumably the oxalic acid was eluted on boiling. The fresh stalk contains about 10–20 mg/100 g of vitamin C **7.1**. The red colour arises from a mixture of cyanidin **7.13** 3-glucoside and 3-rutinoside. The polyhydroxyanthraquinones chrysophanic acid **7.61**, rhein **7.62**, emodin **7.63** and aloemodin **7.64**, together with their monomethyl ethers and glucosides are responsible for some of the astringent taste. The purgative effect of rhubarb is associated with chrysophanic acid **7.61** and aloe-emodin **7.64**.

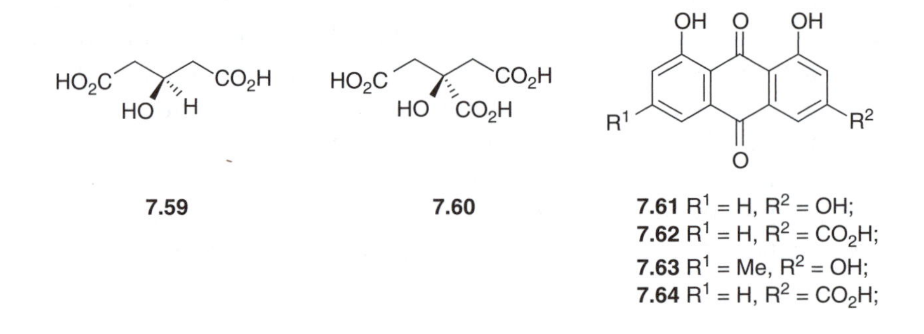

7.59

7.60

7.61 R^1 = H, R^2 = OH;
7.62 R^1 = H, R^2 = CO_2H;
7.63 R^1 = Me, R^2 = OH;
7.64 R^1 = H, R^2 = CO_2H;

7.7 TOMATOES

The tomato (*Lycopersicon esculentum*) belongs to the Solanaceae like the potato. It also produces a range of steroidal alkaloids. The saponin, α-tomatine **7.65**, consists of an aglycone (tomatidine) and a tetrasaccharide (β-lycotetraose) which is constructed from two molecules of glucose and one each of galactose and xylose. It has been found in the leaf, stem, roots, flowers and the green fruit of the tomato plant. The alkaloids provide protection against general microbial infection. The fungus *Fusarium oxysporum* is a widespread soil-borne plant pathogen which causes general vascular wilts. A form of this fungus (*F. lycopersici*), which is a successful pathogen on tomatoes, produces a tomatinase enzyme which cleaves the glycoalkaloid, α-tomatine, into inactive components, tomatidine and β-lycotetraose. Thus the pathogen succeeds in overcoming the plant defences.

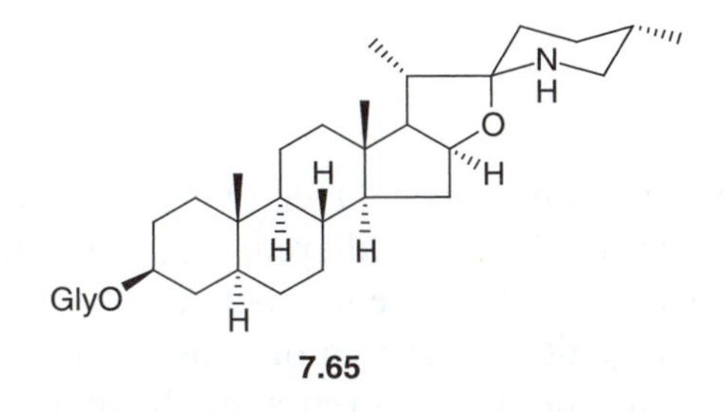

7.65

The tomato contains a wide range of carotenoid pigments which change as the fruit ripens. These include the hydrocarbons phytoene **7.66** and its 15(*Z*) isomer phytofluene, ξ-carotene and lycopene **7.67**, together with some oxygenated derivatives such as phytoene 1,2-oxide. As the fruit ripens other constituents also change. Among the amino acids some, like alanine, decrease but interestingly there is a marked increase by as much as tenfold in the amount of glutamic acid which is present. This can reach as much as 200 mg/100 g fresh weight. Sodium glutamate is a well-known flavour enhancer. The major flavour components of tomatoes are isobutanol, isopentanol, hexanol, 2-methyl-3-butanol, benzaldehyde, hexanal, isopentenal and the esters isopentyl acetate, isopentyl butyrate, isopentyl isovalerate **7.68**, *n*-butyl hexanoate and *n*-hexylhexanoate, β-damascenone **7.69** and β-ionone **7.18**. Tomatoes grown outdoors have been shown to be richer in volatile constituents compared to those grown in the greenhouse or artificially ripened from green samples. There is quite a difference in the volatile constituents of tomato pulp, in which small amounts of dimethylsulfide have been detected.

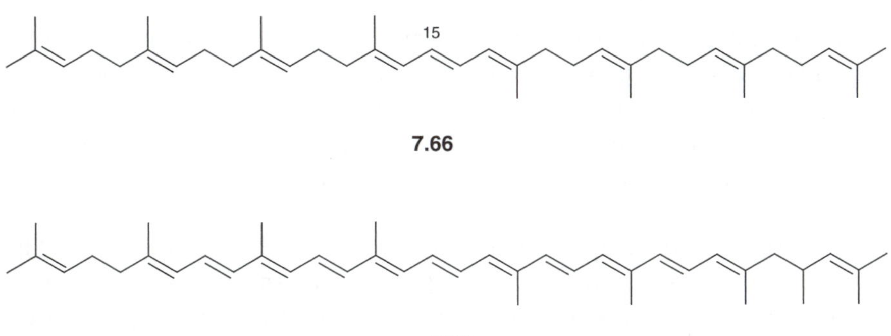

7.66

7.67

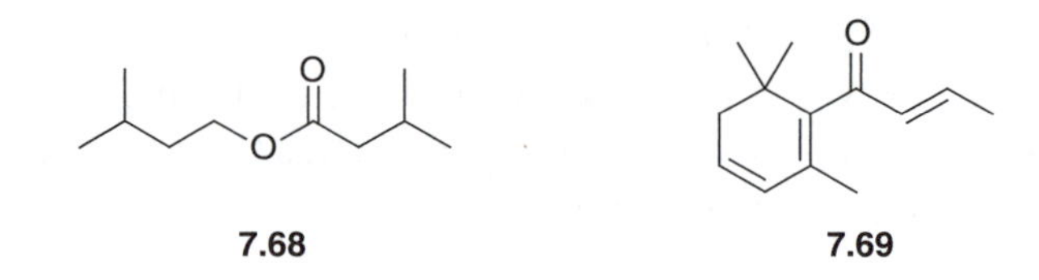

7.68 **7.69**

7.8 FRUIT TREES

Many gardens contain apple trees. Over two hundred volatile components have been identified from different varieties of apples (*Malus domestica*). There are marked differences between the constituents of the aroma of different varieties. Furthermore the profile of these volatile compounds changes as the apple progresses through maturation, harvest and subsequent storage. Rainfall and temperature have also been shown to influence the volatiles from the apples and the apple trees. The major volatile substances from the fruit include esters such as butyl acetate and 2-methylbutyl acetate **7.70** and esters of (+)-2-methylbutanoic acid, ethyl and hexyl butanoate, hexyl acetate, hexenyl acetate, β-damascenone **7.69** and *E*-2-hexenal and *E*-2-hexen-1-ol. The flavour is a combination of these volatile substances with malic **7.59** and citric **7.60** acids and sugars. When apple cells are ruptured the C_6-components increase and become significant components in apple juice. Some apple cultivars accumulate *R*-(+)-octane-1,3-diol **7.71** by degradation of linoleic acid. As with other fruit, the ripening of apples is induced by ethylene. When unripe apples are exposed to the vapour of propionic and butyric acids ripening also occurs.

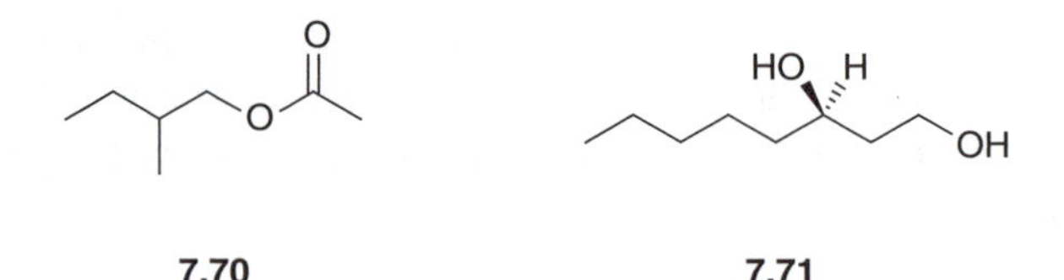

7.70 **7.71**

One of the serious pests of the apple is the codling moth, *Cydia pomonella*. The seasonal variation in the volatiles from the apple actually repel the moth in the early stages of fruit formation up to late May but are attractants in July and August. Bioassays have suggested that benzaldehyde and butyl acetate might contribute to the repellent effect.

'An apple a day keeps the doctor away' so the saying goes. The polyphenolic constituents of the apple play an important role in providing health benefits. Five types of polyphenol, hydroxycinnamates, dihydrochalcones, flavanols, flavones and anthocyanins, together with vitamin C, contribute to the anti-oxidant and radical scavenging properties of apples. The presence of the dihydrochalcone phloretin **7.72** and its 2′-*O*-β-glucopyranoside, phloridzin, are associated with these beneficial effects. The anti-oxidant property of apples is also associated with the presence of oligomeric proanthocyanidins. The proanthocyanidin of apple peel is mainly leucocyanidin **7.73**, whilst the major red pigment is the 3-galactoside of cyanidin **7.13**. These polyphenols give rise to the browning of cut apples.

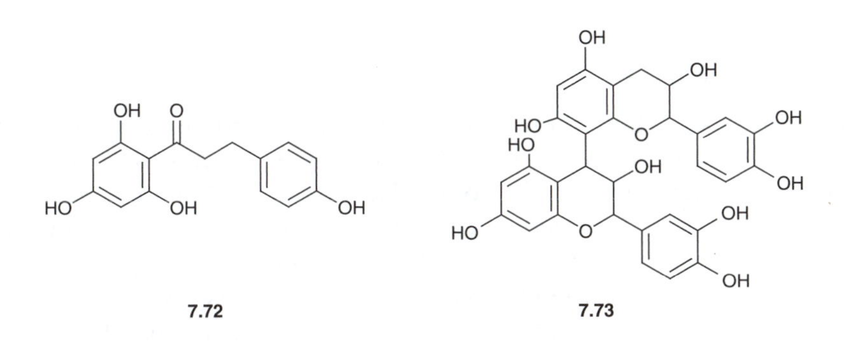

7.72 **7.73**

A number of fungi are found growing on apples, particularly in storage. A pink rot is often *Trichothecium roseum*. This fungus produces two sets of metabolite, a sesquiterpene, trichothecin **7.74**, which imparts a bitter taste to the apple, and the diterpenoid rosane lactones such as rosenonolactone **7.75**. A more serious problem is caused by a grey-bluish mould, *Penicillium expansum*, which gives a soft brown rot often from the core of stored apples. This fungus produces patulin **7.76** which can contaminate apple juice and other apple products. Patulin **7.76** is mutagenic and a maximum residual level of 50 μg/kg has been recommended. A brown rot with grey or yellow pustules of conidia may be caused by a *Sclerotinia* species, *e.g. S. fructigens*. The infected fruit gradually shrivel as their structure is broken down and they dehydrate.

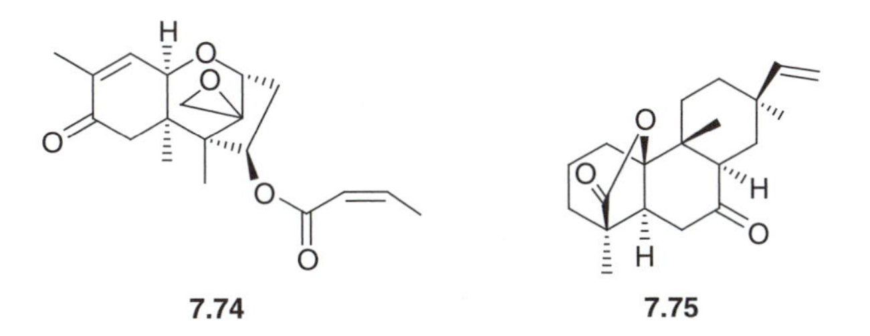

7.74 **7.75**

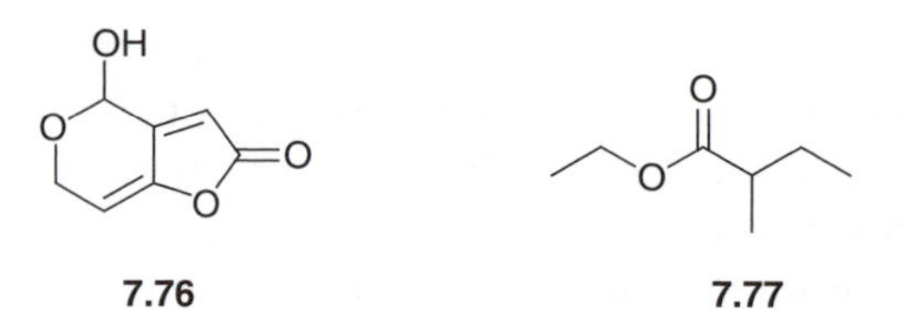

7.76 **7.77**

The flavour of pears (*Pyrus communis*) is associated with the presence of the fruit esters butyl and hexyl acetate, ethyl butanoate and ethyl (+)-2-methylbutanoate **7.77**, together with the methyl and ethyl esters of (2*E*,4*Z*)-2,4-decadienoic acid.

7.9 SOFT FRUIT

The garden strawberry (*Fragaria* x *ananassa*) as opposed to the alpine or wild strawberry, is a cross between several species. The deep red colour of strawberries arises from pelargonidin **7.78** and cyanidin **7.13** 3-glucosides, together with lesser amounts of the 3-arabinoside, the 3-(6″-malonylglucoside) and 3-(6″-rhamnosylglucoside) of pelargonidin. An anthocyanin with the unusual aglycone 5-carboxypyranopelargonidin **7.80** has also been detected. The fresh fruit contains glucose and fructose, and there is quite a high vitamin C content (c. 70–80 mg/100 g). The high water content and rather low pectin and acid (citric and malic acids) content can pose difficulties in making jam.

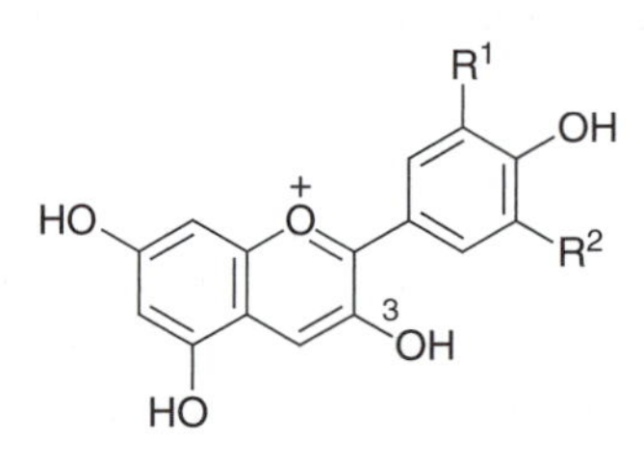

7.78 $R^1 = R^2 = H$
7.79 $R^1 = R^2 = OH$

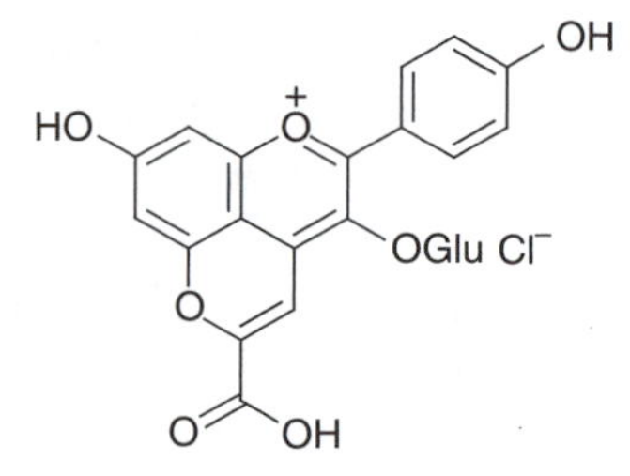

7.80

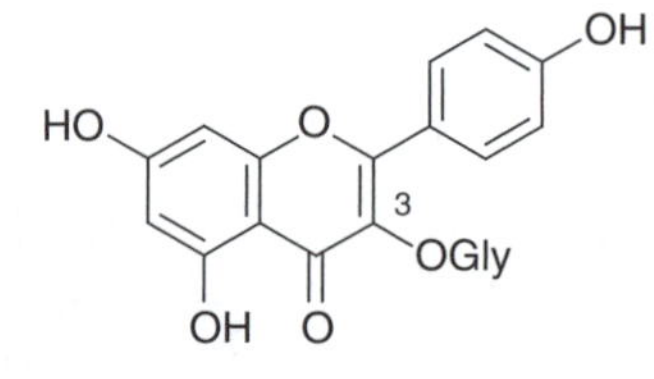

7.81

The anti-oxidant properties of strawberries have been associated with their anthocyanin content. Many drugs are metabolized in the liver by cytochrome P_{450} (CYP) enzymes. The most abundant of these haem mono-oxygenases is CYP3A4 which is responsible for the oxidation of about 50% of clinically used drugs. Inhibition of this enzyme by, for example, grapefruit juice, reduces the first pass loss and hence alters the bioavailability of drugs. Kaempferol-3-β-D-(6-*O-cis* and *trans-p*-coumaroyl)glucopyranosides **7.81** which inhibit this enzyme have been isolated from strawberries.

Studies on the hormonal control of berry enlargement have implicated auxins and gibberellins in these developmental processes. Studies on the endogenous gibberellins have revealed two sequences of gibberellins belonging to the 13-hydroxylated series and containing additional hydroxyl groups at C-3 and C-12, *e.g.* **7.82**.

The blackcurrant (*Ribes nigra*) is widely grown as a soft fruit. A characteristic feline odour is often observed both on pruning bushes and on harvesting the fruit. This arises from 4-methoxy-2-methylbutane-2-thiol **7.83**. Another compound with a feline odour is 4-mercapto-4-methylpentan-2-one which is found in the box tree. The other major components of blackcurrant flavour are fruit esters such as ethyl butanoate, the monoterpenes, 1,8-cineole, linalool and geraniol, 4-methoxyacetophenone and 2,3-butanedione. Many blackcurrant drinks utilize buchu leaf oil to supplement the blackcurrant flavour. There is quite a high vitamin C content (72–191 mg/100 g) in blackcurrants. The colour of blackcurrants is associated with the presence of delphinidin **7.79** and cyanidin **7.13** glycosides. The related redcurrant has a lower vitamin C content and utilizes mainly cyanidin glycosides as the pigments.

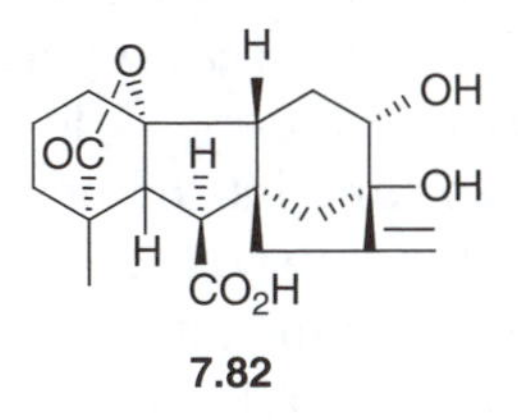

7.82

The flavour of raspberries (*Rubus ideaus*) is based on raspberry ketone, 4-(4-hydroxyphenyl)-2-butanone **7.84**. The terpene derivatives, α- and β-ionones **7.18** and β-damascenone **7.69**, geraniol and linalool, together with *Z*-3-hexen-1-ol and citric acid, also make a contribution to this. Their colour is based on cyanidin **7.13** glycosides. The anti-oxidant properties of the anthocyanin content of blackberries, blackcurrant, blueberries and raspberries has been linked to their beneficial health effects. There are also reports concerning the reduction of cancers. Since the growth rate of cancer cells generally outpaces that of normal cells, the induction of apoptosis or 'programmed cell death' is a potentially useful method of cancer suppression. Berry extracts and in particular the anthocyanin fraction from strawberries and a black form of raspberry (distinct from blackberry) have been shown to exhibit this pro-apoptopic activity.

The constituents of the grape (*Vitis vinifera*) and the wines produced from the various cultivars have been thoroughly investigated. It has been estimated that over 800 compounds have been detected in the grape and various wines. Tartaric acid **7.85** is the major fruit acid which is present, together with smaller amounts of malic **7.59** and citric **7.60** acids. The flavour of the grape includes a subtle blend of geraniol, linalool (both monoterpenes), β-damascenone **7.69** and vitispirane **7.86** (both norsesquiterpenoids). The fruit esters such as ethyl 3-methylbutanoate also make a contribution. There is a high sugar content. Proline and arginine are the predominant amino acids that are present together with smaller amounts of glutamic acid, glutamine and alanine. The anthocyanin colouring matter of the skin depends on the variety but includes the 3-glucosides of the methyl ethers malvidin and peonidin, together with delphinidin **7.79** and cyanidin **7.13**.

Two constituents of the skin are the hydroxystilbenes resveratrol **7.87** and its oligomeric derivative viniferin **7.88**. Resveratrol is a phytoalexin and possesses anti-microbial properties. It is found in red wine and is a powerful anti-oxidant which confers considerable health benefits. The French have a similar consumption of dietary fat but a much lower incidence of coronary heart disease when compared to people in the UK. This difference was traced to the higher consumption of red wine by the French and eventually to the effects of resveratrol **7.87** arising from the red grape skins. The cardioprotective effect of resveratrol was associated with its anti-oxidant activity. By scavenging reactive oxygen species, resveratrol inhibits the peroxidation of low density lipoprotein and platelet aggregation. It also inhibits the enzyme 5-lipoxygenase which mediates a key step in the biosynthesis of the leukotrienes from arachidonic acid. This confers some anti-inflammatory activity. Resveratrol also possesses some tumour inhibitory activity. Its presence provides a very good excuse for enjoying a glass of red wine along with the products of the garden!

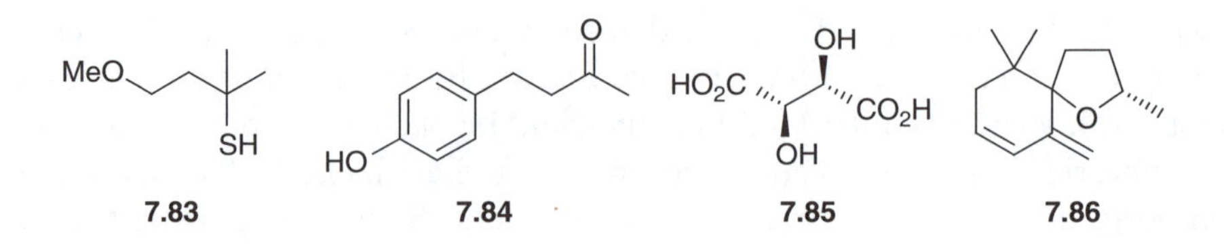

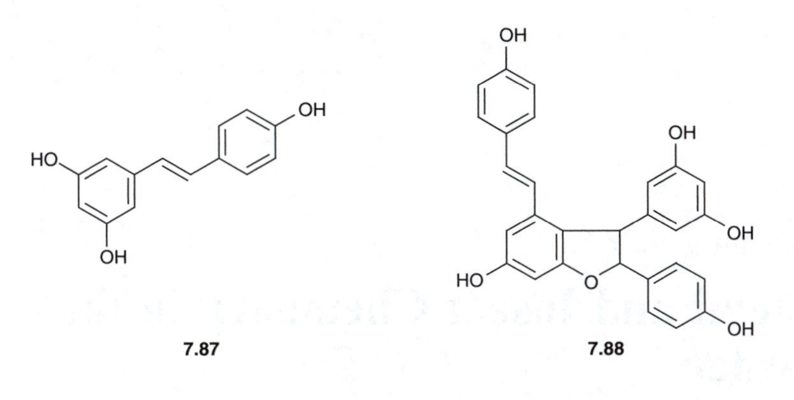
OH
HO
OH
7.87
OH
OH
OH
HO
O
OH
7.88

CHAPTER 8

Fungal and Insect Chemistry in the Garden

The gardener may use commercial fungicides and insecticides in order to control fungal disease and insect pests in the garden. However in nature the interaction between an infective organism or an insect pest and a plant is mediated by natural products. Many aspects of this chemical ecology have been examined over the past fifty years. There is an immense structural diversity in the compounds that are involved and in many cases the biological responses are to mixtures of compounds. In this chapter we shall consider just a few of the many interactions that have been studied and which involve garden flowers and vegetables and their relationships with fungi and insects.

8.1 MICROBIAL INTERACTIONS

Natural products are involved in many of the defence mechanisms of plants against attack by phytopathogenic micro-organisms. Some anti-bacterial and anti-fungal natural products are constitutive and are present in anticipation of attack. They are sometimes called phytoanti-cipins. The formation of others may be induced as a response to stress or to microbial attack. These are phytoalexins. The possibility that compounds of this type existed was suggested in 1940 in studies on the interaction between the fungus *Phytophthora infestans* and the potato (*Solanum tuberosum*). In 1960 the first phytoalexin was isolated and its structure established. This was pisatin **8.1** which was obtained from the pea, *Pisum sativum*. Since then many compounds with diverse structures have been detected as phytoalexins in different plants. Phytoalexins, such as pisatin, were active against a range of fungi many of which were not pathogens on peas. However pisatin was less toxic to a pea pathogen, *Aschochyta pisi*. It became apparent that an organism which was a virulent pathogen on a particular plant, was tolerant to the phytoalexins produced by the plant and, in many cases, it was able to detoxify them. The fungus *A. pisi* and a number of other pea pathogens can metabolize pisatin **8.1** to the less active (+)-6a-hydroxymaackain **8.2**. Phytoalexins provide a general rather than a specific defence mechanism against microbial attack.

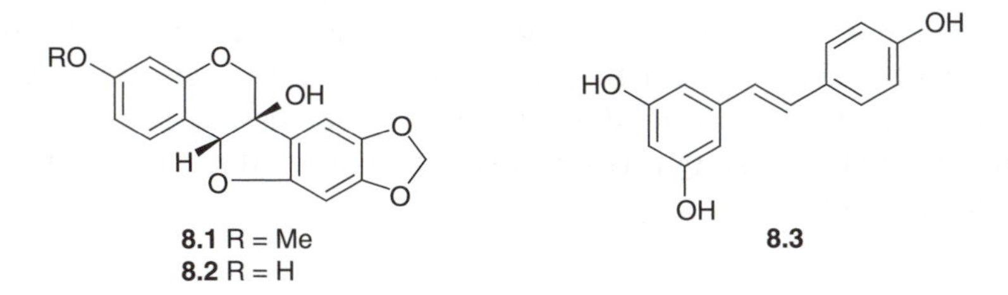

8.1 R = Me
8.2 R = H

8.3

Disease resistance can be enhanced by importing a foreign phytoalexin into a particular plant. Thus when the enzyme system, stilbene synthase, was transferred from grapes (*Vitis vinifera*) to tobacco plants (*Nicotiana tabacum*) the latter then made the phytoalexin resveratrol **8.3** and showed an increase in resistance to a strain of the fungus *Botrytis cinerea* which was normally a pathogen on tobacco. However other strains of *Botrytis cinerea* can detoxify resveratrol. The enzymatic detoxification of phytoalexins is part of the natural product warfare in the garden.

Another aspect is revealed by considering the mechanism of attack by the pathogenic organisms. The arrival of a pathogenic organisms on a plant cell surface may be recognized by the glycoprotein interaction between the plant and the micro-organism. This then stimulates phytoalexin production by the plant. On reaching the plant the micro-organism produces extra-cellular enzymes that destroy plant tissue and allow the phytotoxic metabolites to enter the plant and weaken or kill it. This entry into the plant may be facilitated by mechanical or insect damage to the plant, features with which the gardener may well be aware. Hence the need to protect plant wounds after pruning.

Many of these general points are illustrated by the way in which a common plant pathogen, *Botrytis cinerea*, attacks plants. This fungus is a grey powdery mould that is found on garden plants such as strawberries, lettuces, grapes and tomatoes. Although it is normally a pathogen on grapes, when it infects ripe grapes late in the season, it facilitates the evaporation of water from the pulp and produces a dehydrated grape with a higher sugar level. Since the fermentation may stop before all the sugar is used, these grapes produce a sweet 'spatlese' wine. *B. cinerea* is then known as the 'noble mould' (Edelfaule). The laccase produced by the fungus also mediates the oxidation of the anthocyanins and phenols in the grape to generate the golden-brown colour of the botrytized wine.

The relatives *B. fabae, B. allii* and *B. tulipae* are pathogens of beans, onions and tulips respectively. *B. cinerea* produces a family of sesquiterpenoid botryane metabolites of which the most phytotoxic is the dialdehyde botrydial **8.4**. Some strains produce another family of phytotoxins known as the botcinolides (*e.g.* botcinic acid **8.5**). When *B. cinerea* has infected a plant, the botryane metabolites have been found within the plant. The fungus also produces an extracellular laccase which facilitates the destruction of plant tissue. The formation of this laccase by the fungus is stimulated by plant natural products such as gallic acid **8.6**.

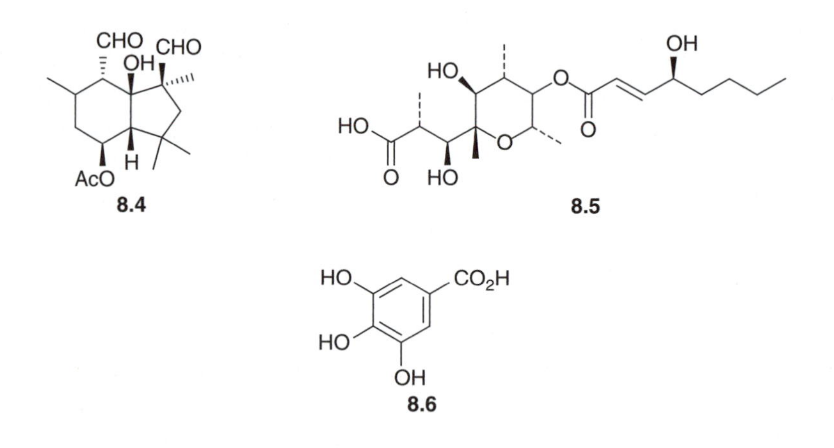

When broad beans were infected with *Botrytis cinerea*, they responded by producing their phytoalexin, wyerone acid **8.7**, in amounts ranging from 3.5–30 μg/g of fresh tissue. These levels were consistent with the anti-fungal activity of the phytoalexin. This phytoalexin was named after Wye College, now part of Imperial College, where it was first isolated. Its structure is quite different from that of the grape phytoalexin.

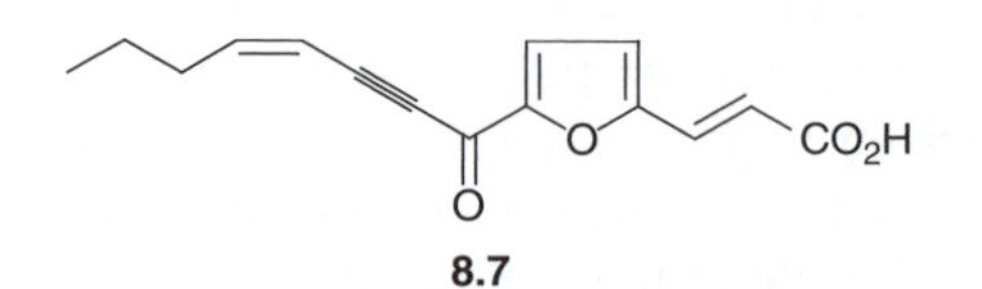

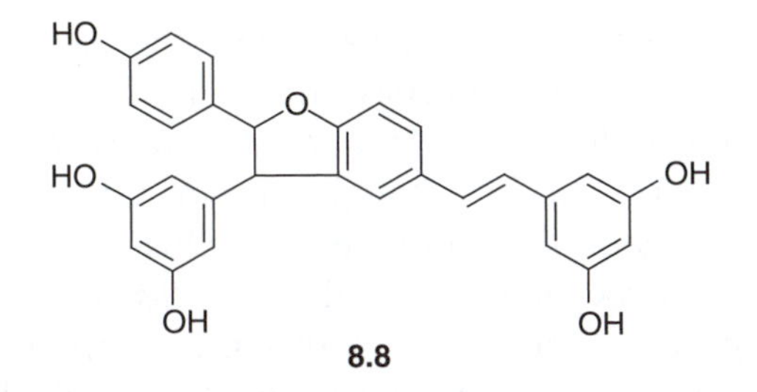

8.8

When *Botrytis cinerea* acts as a pathogen of grapes they produce the hydroxystilbene resveratrol **8.3** as a phytoalexin. However the virulent phytopathogenic strains of *B. cinerea* can metabolize the resveratrol to an inactive dimer **8.8** in a reaction that is possibly mediated by a stilbene oxidase.

Use can be made of this infective sequence in designing selective antifungal agents. Thus many biosyntheses have a feedback regulatory control by one of the products of the biosynthetic pathway. This prevents the fungus producing excessive amounts of a toxic metabolite and suffering auto-toxic effects. Incubation of the fungus with a non-phytotoxic analogue of the phytotoxic metabolite can sometimes not only diminish natural phytotoxin production but also restrict microbial growth, an effect which has been observed with *B. cinerea*. This artificial regulatory control of the biosynthesis of the phytotoxins allows the plant defences to then dominate the fungal attack.

The fate of the phytoalexins produced by several garden plants from the Solanaceae has been studied. Quite simple changes by virulent pathogens lead to detoxification. Thus the phytoalexin from pepper (*Capsicum annuum*) capsidiol **8.9** is oxidized to the unsaturated ketone capsenone by strains of *B. cinerea* and *Fusarium oxysporum*, whilst the potato phytoalexins lubimin **8.10** and rishitin **8.11** are detoxified by the potato pathogen *Gibberella pulicaria* by epoxidation.

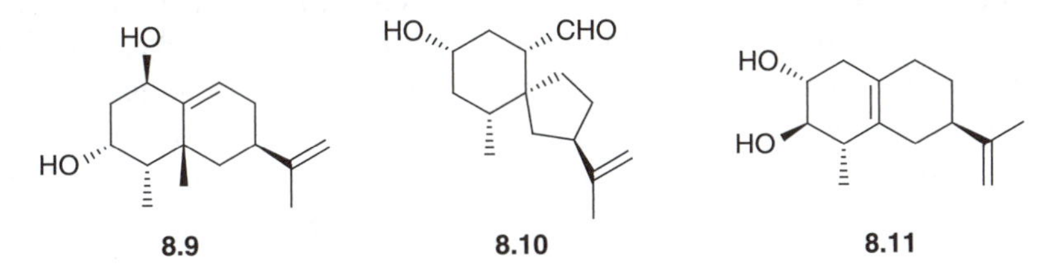

8.9 **8.10** **8.11**

Dutch elm disease reveals the intervention of another parameter in the attack of a fungus on a plant, that of an insect vector, a bark beetle. The elm produces the feeding attractants catechin 7-xyloside **8.12** and a triterpenoid ester (lupeyl cerotate) to which the bark beetle *Scolytus multistriatus* responds. The beetle brings with it the spores of the fungus *Ceratocystis ulmi*. The fungus *C. ulmi*, has two types of toxin, a glycoprotein and some phenolic metabolites, *e.g.* **8.13**. These cause the wilting and death of the tree. Beetles are deterred from feeding on trees which have already been infected by another fungus, *Phomopsis oblonga*. This organism produces deterrent compounds including 5-methoxymellein **8.14**. Since the aggregation pheromones (δ-multistriatin **8.15**, 4-methyl-3-heptanol **8.16** and α-cubebene **8.17**) of the *Scolytus* bark beetle are known, it is possible to produce traps laced with these compounds and estimate the number of beetles in the area and hence the potential for infection.

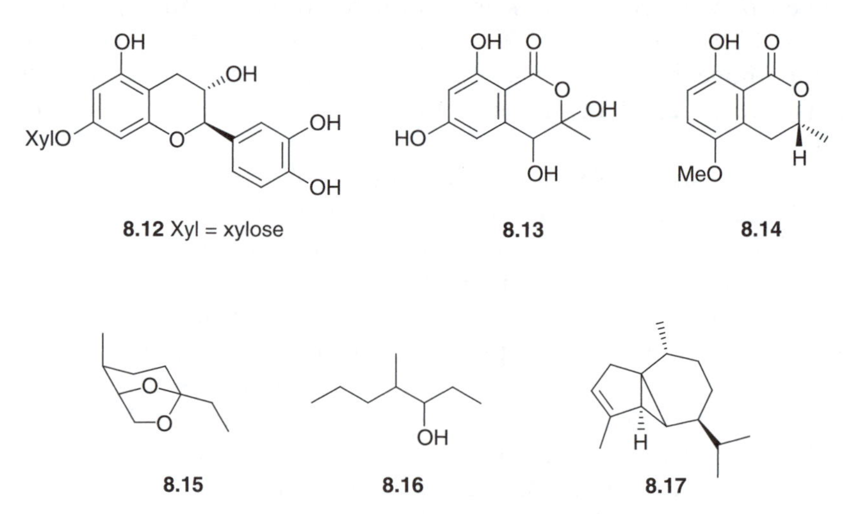

A number of fungi attack the grass in the lawn. Brown patches can be produced by *Fusarium culmorum*. This particular organism produces three groups of sesquiterpenoid fungal metabolite exemplified by cyclonerotriol **8.18**, culmorin **8.19** and isotrichodermin **8.20**. The latter is a trichothecene mycotoxin and may be responsible for the phytotoxic effect. The *Fusaria* are widespread soil-borne plant pathogens. The more highly hydroxylated trichothecenes that are produced by the *Fusaria* are serious mycotoxins. For example the 'snow fungus', *F. nivale*, so called because it will grow on fescue grass under snow, produces nivalenol **8.21** and its relatives. When transferred to the ankles and feet of grazing cattle, it produces necrotic skin lesions know as 'fescue foot' which can cripple the animals. Other *Fusaria* are spoilage organisms on corn and produce toxic metabolites with highly descriptive names such as vomitoxin **8.22**.

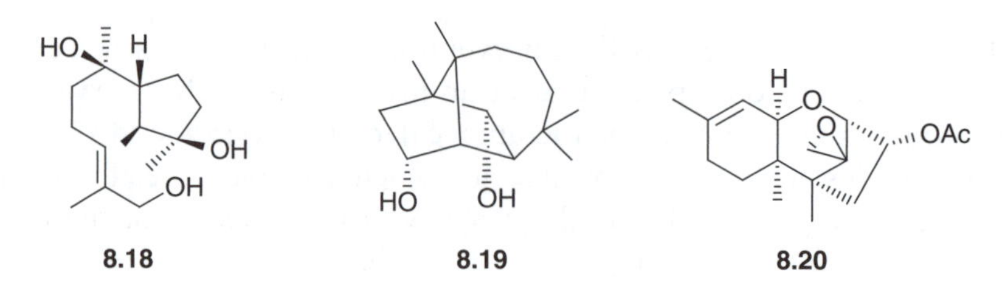

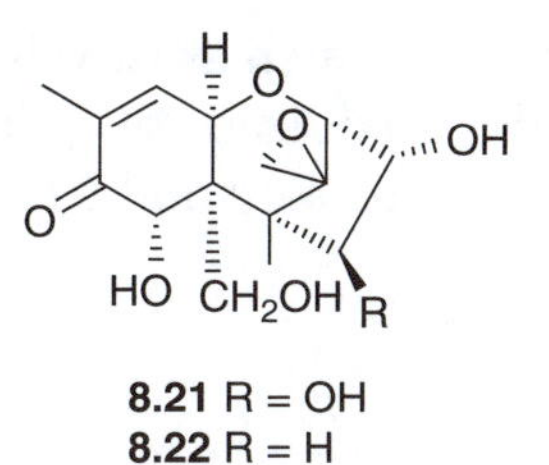

8.21 R = OH
8.22 R = H

The 'fairy ring' fungus found in the lawn, *Marasmius oreades*, is a typical Basidiomycete, producing a fruiting body from underground mycelium. The fact that the soil under the lawn remains relatively undisturbed allows the mycelium to grow outwards from a central infection and, when sporulation occurs, the fruiting bodies are then produced at the periphery of the colony, forming the ring. These rings have a zone of stimulated grass growth and an inner weakened or even dead zone. The zone of stimulated growth arises from the mycelial release of nutrients for the grass from the humus. Many of the Basidiomycetes are noted for the terpenes and polyacetylenes that they produce. *M. oreades* is no exception, producing a family of drimane sesquiterpenoids of differing oxidation pattern and exemplified by marasmone **8.23** together with a phytotoxic polyacetylene, agrocybin **8.24**. The latter may be responsible for killing the grass.

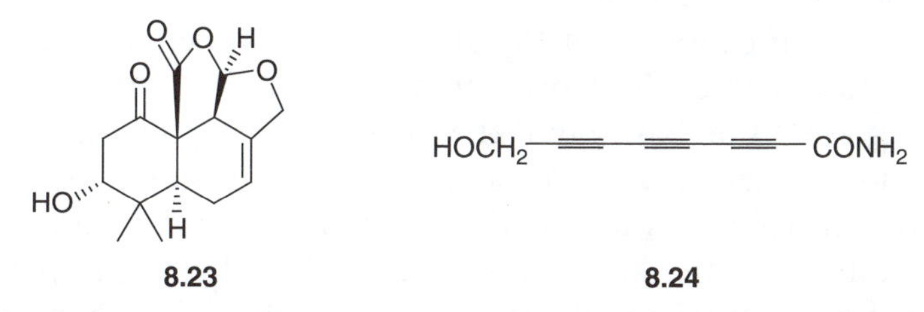

8.23 **8.24**

Fungi of the genus *Lactarius* grow on sandy soils under deciduous trees and their fruiting bodies sometimes appear in the autumn in the garden. A characteristic of the genus *Lactarius* (milk caps) which are also Basidiomycetes, is that their fruiting bodies contain a latex, droplets of which can be seen when the cap is cut. The colour changes of the latex as it undergoes aerial oxidation can have taxonomic significance. In the case of *L. delicious* the colour changes to the latex are associated with the formation of azulenes such as lactaroviolin **8.25** and lactarofulvene. Young undamaged specimens contain the dihydroazulene alcohol **8.26** as its stearic acid ester, which undergoes enzymatic hydrolysis and autoxidation as the tissue is damaged. Other *Lactarius* species (*e.g. L. vellereus*) which are resistant to attack by insects, snails and small mammals have a peppery taste. This defence mechanism has been traced to the enzymatic conversion of the stearic acid ester of velutinal **8.27** to the pungent tasting dialdehydes such as isovelleral **8.28** and velleral **8.29**. Some of the more highly oxidized sesquiterpenoid metabolites of *Marasmius oreades* contain a masked dialdehyde and this might provide a similar protection.

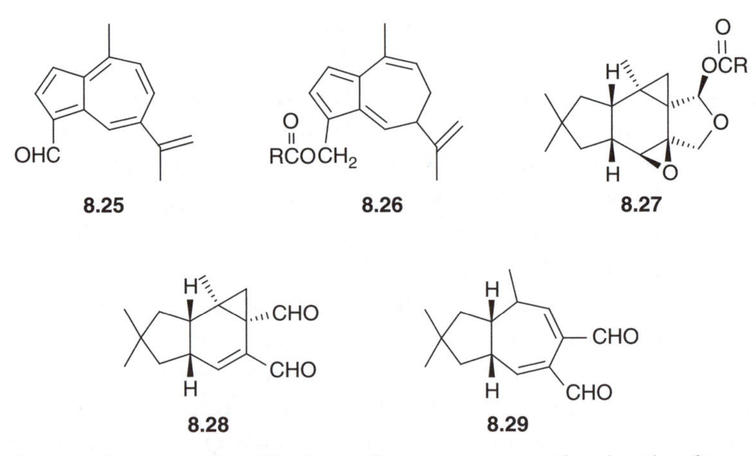

The honey fungus, *Armillaria mellea*, can cause the death of trees such as apples, flowering cherries, willows, birch and walnuts. The fungus gets its name from the honey-coloured fruiting bodies which appear around the trees. The fungus is also known as the boot-lace fungus because of the appearance of the rhizomorphs which spread from the roots of infected trees. The mycelium spreads as a white sheet sandwiched between the bark and the wood of the tree, blocking the translocation of water and nutrients to the leaves. Hence the leaves which are deprived of water and nutrients rapidly wilt and die. The fungus produces a number of sesquiterpenoid phytotoxic metabolites exemplified by armillyl orsellinate **8.30** and melleolide **8.31**.

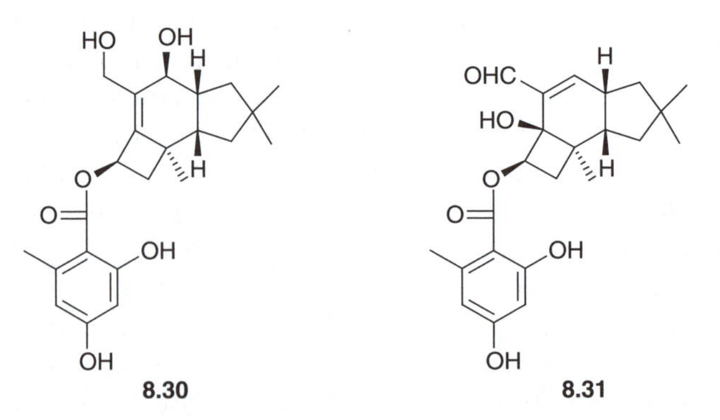

The bracket fungi found on trees such as the birch are also Basidiomycetes and produce triterpenes. For example *Piptoporus* (*Polyporus*)

betulinus produces polyporenic acid **8.32**. The dark coloured pigments of these fungi include diphenylbenzoquinones such as polyporic acid **8.33**.

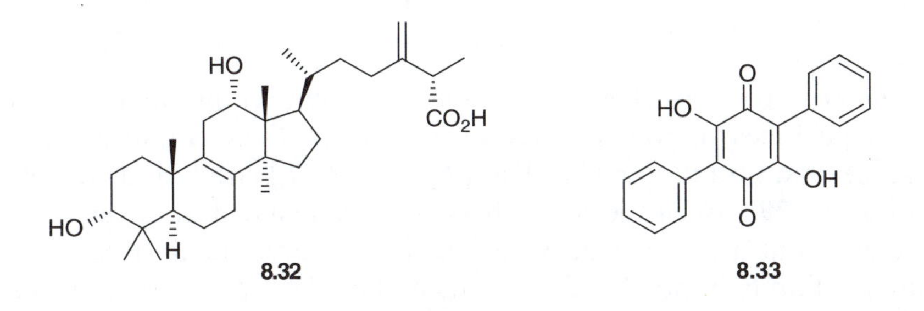

8.32 **8.33**

8.2 LICHENS

Many surfaces within the garden attract lichens. These organisms are a symbiotic growth between an alga or cyanobacterium and a fungus. The photosynthetic partner produces sugars for the fungus. However the natural products which they produce, particularly the colouring matters, are a characteristic of the lichen rather than the individual components. Lichens draw their mineral nutrients from the water running over the surface to which they are attached. Consequently their presence is often used as a measure of environmental pollution. They are more widespread in the damper parts of the country.

Lecanora species such as *L. campestris* and *L. dispersa* are quite widespread forming grey or greyish white crusts on walls and concrete. *L. dispersa* is quite tolerant of pollution and may be found in cities. *Psilolechia lucida* is a greenish-yellow powdery lichen found on moist brickwork. Typical natural products found in these lichens include a series of esters of polyketide aromatic acids with related phenols. These are known as the depsides and are exemplified by lecanoric acid **8.34** which contains two molecules of orsellinic acid. A series of xanthones, *e.g.* lichexanthone **8.35**, have also been isolated. Many of these contain chlorine and may provide anti-microbial protection for the lichen. There are also some metabolites with components derived from the tricarboxylic acid cycle such as lichesterinic acid **8.36**.

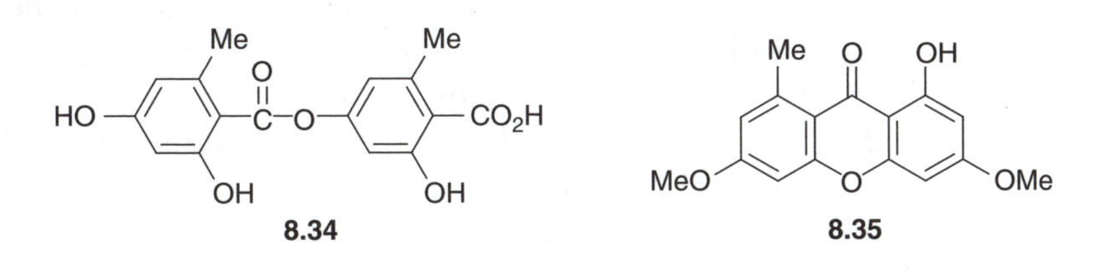

8.34 **8.35**

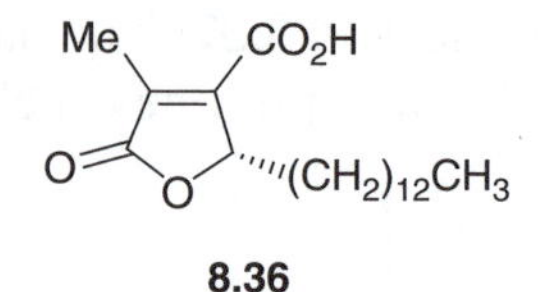

8.36

Xanthoria species such as *X. parietina* form quite common deep yellow or orange lobes on walls. They owe some of their colouring to the presence of anthraquinones such as physcion **8.37**, parietinic acid **8.38** or xanthorin **8.39**. *Usnea* species such as *U. florida* have dark green or grey structures and are found growing on trees in unpolluted areas. These contain a family of polyketides in which phenol coupling reactions have played an important role in their biosynthesis. These are exemplified by usnic acid **8.40** which is formed by the coupling of two molecules of methyl phloroacetophenone **8.41**.

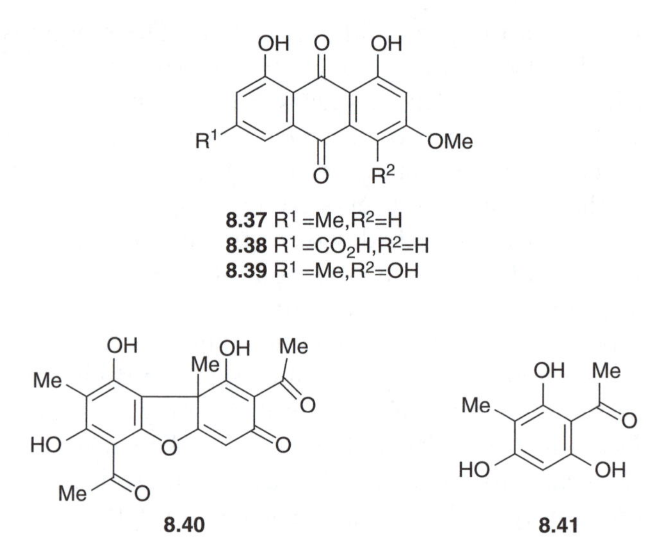

8.3 MYCORRHIZAL AND ENDOPHYTIC ORGANISMS

Not all fungal–plant interactions are deleterious. Mycorrhizal fungi which grow around the roots of plants mobilize nutrients and minerals for uptake by the plant, whilst the nitrogen fixing bacteria of the legumes have an extremely important role in the successful development of these plants. When moving a plant it is wise to move some of the surrounding soil as well, in order to take the mycorrhizal organisms. Recently deoxystrigol **8.42**, a relative of strigol which is the stimulant of the parasitic plants *Striga* and *Orobanche* has been found to be produced as a germination stimulant for mycorrhizal organisms.

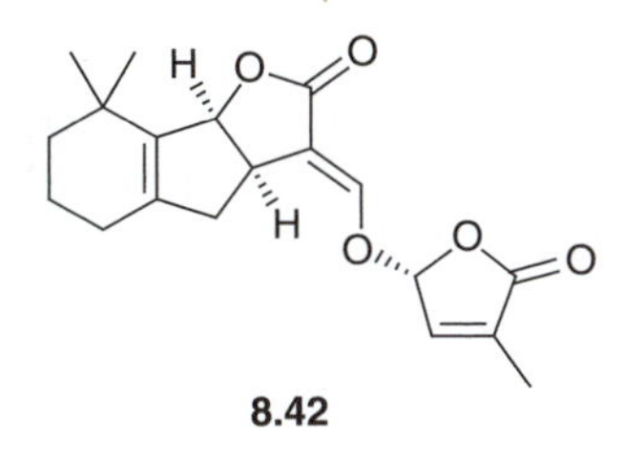

8.42

Another relationship between fungi and plants is that of the endophytic organisms. Endophytic micro-organisms grow within the plant. Their relationship with their host plant may be symbiotic or that of a latent phytopathogen, only exerting their biological effect on a stressed plant. There are instances in which the endophyte produces a plant hormone such as indolyl-3-acetic acid, stimulating the growth of the plant. *Acremonium* (*Neotyphodium*) species have been found within rye grass (*Lolium perenne*). The amine peramine **8.43** and some ergot alkaloids such as ergovaline have been isolated. The latter were also isolated from a fescue grass and may have been responsible for livestock toxicosis. Some tremorgenic indole alkaloids have been obtained from an endophyte infected rye grass and may also be the cause of ryegrass staggers in livestock. The lolines, *e.g.* **8.44**, are quite powerful insectides which are produced in endophyte infected grasses. It has been reported that some endophytes isolated from *Taxus* sp. (yew) are capable of producing the anti-cancer drug taxol®.

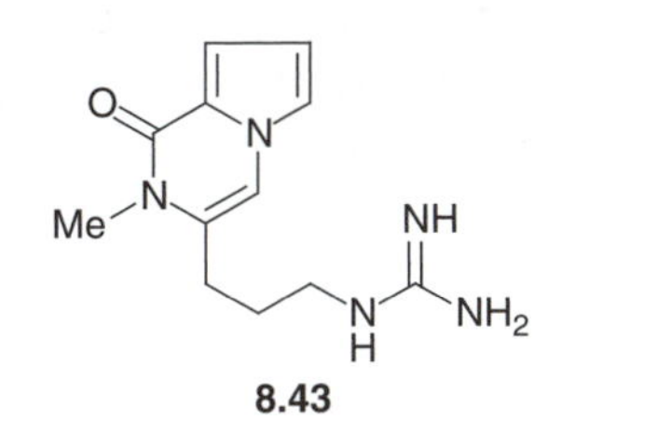

8.43

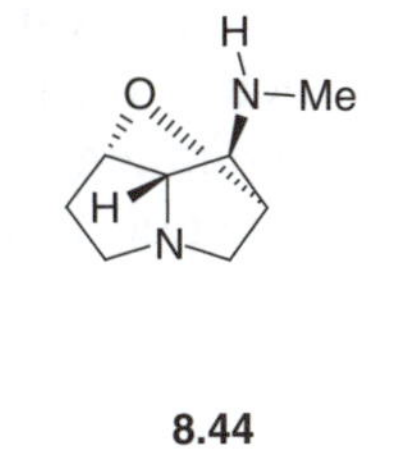

8.44

8.4 INTERACTIONS BETWEEN FUNGI

Interactions mediated by natural products take place not just between fungi and plants but also between fungi. Some strains of the soil fungus *Trichoderma viride* (*Gliocladium roseum*) produce the powerful anti-fungal agents gliotoxin **8.45** and viridin **8.46**. These anti-fungal compounds prevent the development of other fungi, giving a competitive advantage to the *Trichoderma*.

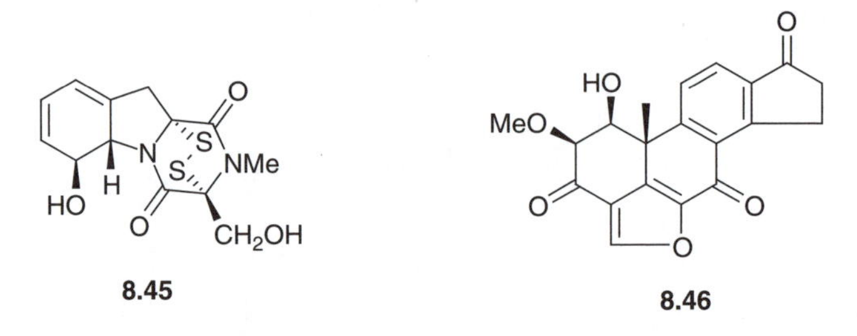

8.45 **8.46**

The chemistry of the interaction between these fungi has been put to use. *Trichoderma* species such as *T. harzianum* are very common competitive micro-organisms in the soil. They exert their dominance over other micro-organisms in three ways. Firstly *T. harzianum* produces a volatile metabolite 6-pentylpyrone **8.47** which permeates through the surrounding soil and inhibits the germination of other fungi. 6-Pentylpyrone has a coconut odour. Along with methylisoborneol **8.48** and geosmin **8.49** which are produced by *Streptomycetes*, it also contributes to the 'earthy' smell of some water from wells. The second competitive group of natural products produced by *T. harzianum* are close-contact anti-fungal agents exemplified by harzianopyridone **8.50** and gliotoxin **8.45**. The third stage involves the production of extracellular enzymes which destroy the fungal cell wall of other organisms. *Trichoderma* species have been used as biocontrol agents to minimize fungal infection of crops, whilst synthetic 6-pentylpyrone when allowed to diffuse through the soil around young seedlings, prevents the ravages of damping-off organisms such as *Pythium ultimum* and *Rhizoctonia solani*. 6-Pentylpyrone has also been detected in the skin of some fruit such as the peach where it may have an anti-fungal role. Its use has been recommended as a fumigant to prevent microbial spoilage of fruit in storage and transit.

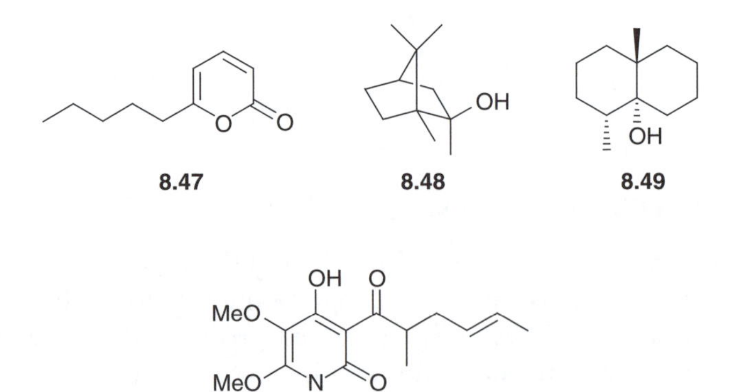

8.47 **8.48** **8.49**

8.50

8.5 INSECT CHEMISTRY IN THE GARDEN

Flowers attract pollinating insects by visual and olefactory cues. There are a large number of compounds which are produced by flowers and blended into complex mixtures to guide and stimulate pollinating insects. These compounds are mainly benzenoid aromatics such as methyl benzoate, benzyl alcohol, benzyl acetate, benzaldehyde, 2-phenylethanol and methyl salicylate. The phenylpropanoid substance eugenol **8.51** is also quite common. The monoterpenoids geraniol **8.52**, nerol **8.53**, linalool **8.54**, myrcene **8.55**, α-pinene **8.56**, β-ocimene **8.57** and 1,8-cineole **8.58**, together with the sesquiterpenoids caryophyllene **8.59** and α-farnesene **8.60**, are also widespread.

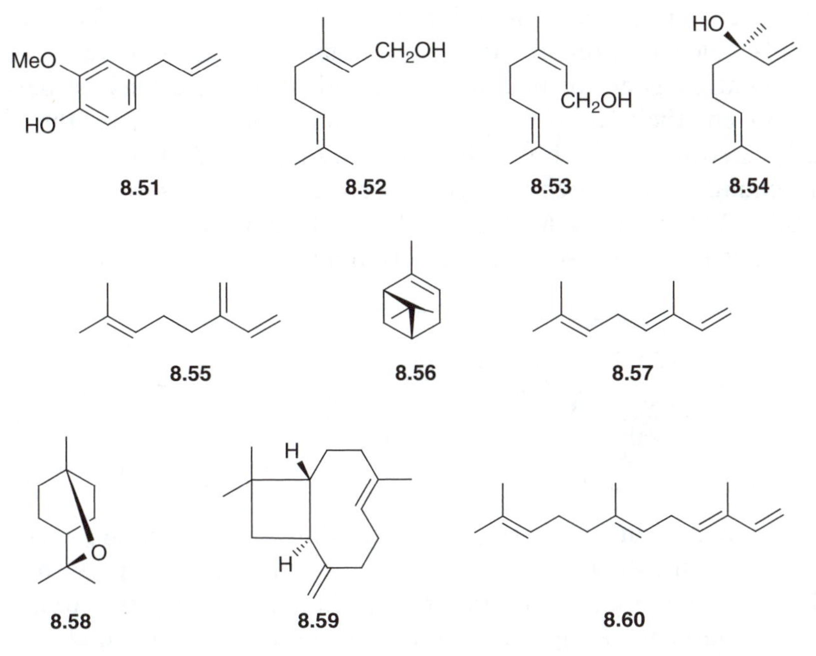

Pollinating insects such as bees are rewarded by the presence of nectar in the flowers. This sugar solution acts as a food source. Honey bees, *Apis mellifera*, can be trained to recognize the volatile cues from particular plants by offering them a reward of sucrose as a pseudo-nectar and an olefactory stimulus of plant volatiles. The role of various attractants have been established in this way. The bee stores the sugar from nectar as honey in the hive. The sucrose content of the original nectar is reduced by the presence of the enzyme invertase. Honey contains 8–25% water, 30–42% glucose and 23–39% fructose, and smaller amounts of sucrose. There are also small amounts of acid which can minimize bacterial fermentation. Honey has a pH 3.8–4.3 and may contain traces of formic, acetic, citric and malic acids. Honey may contain small amounts of pollen and volatile constituents of the plant which the bee has visited. Pollen can provide a protein source for the bee. Occasionally plant toxins, such as the diterpenoid grayantoxin acetylandromedol **8.61** from rhododenrons, may accumulate in the honey with serious consequences for those who eat this honey. The poisoning of Greek soldiers by honey derived from bees feeding on *Rhododendron* species was recorded by Xenophon writing in 400 BC and the effects of 'mad honey' (grayanotoxin intoxication) are still reported today particularly in the eastern Black Sea region of Turkey where *R. ponticum* is widespread.

Beeswax from the honeycomb is a mixture of straight chain C_{25}–C_{33} hydrocarbons possessing an odd number of carbon atoms. The C_{24}–C_{32} alcohols which are present have an even number of carbon atoms. Some are present as esters of palmitic acid. Although these may be derived from pollen, their uniformity suggests some metabolism by the bees. Many functions of the hive are chemically regulated. Thus the mandibular glands of the queen honeybees secrete queen bee substance, *trans*-9-oxodec-2-enoic acid **8.62**, which inhibits the development of ovaries in the worker bees and prevents the development of further queens.

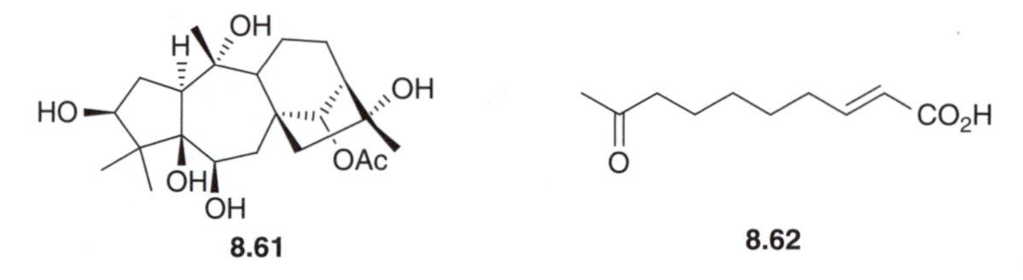

Insects make considerable use of signalling substances, semiochemicals, in regulating their actions. Many examples have been established with insects that are pests of major crops and a number apply to insects that are found in the garden. Thus the observation that male moths could be attracted to flutter around a cage which held a female moth was made in the nineteenth century. The implication that volatile chemicals are attractants led eventually to the isolation of the pheromones. In many cases the responses are to mixtures of compounds which are active at the submicrogram level. Insect pheromones have a sexual function or act as aggregation pheromones. Others are trail or alarm pheromones and are particularly important with social insects such as ants. Many of the sex attractants, particularly of the butterflies and moths, are relatively simple alkenes, alkenyl alcohols and esters. The response of a particular insect to these is dependent on the position and geometry of the double bond and on the composition of the mixtures. Thus the attractant of the house fly, *Musca domestica*, is muscalure, *Z*-9-tricosene. The role of aggregation pheromones in communication between bark-boring elm beetles, *Scolytus multistriatus*, has already been mentioned in the context of the spread of the fungal Dutch elm disease. Other bark beetles including *Ips* species, modify terpenes such as myrcene **8.55** that are found in the bark to produce, for example, the aggregation pheromone ipsdienol **8.63**. There is evidence to suggest that there is a synergistic interaction between some beetle aggregation pheromones such as ipsdienol and plant constituents such as α-pinene. Overcrowding can lead to the production of pheromones to diminish the attractiveness of a site. Verbenone, an unsaturated ketone derived from α-pinene **8.56**, fulfils this role. Although it is photolabile, its use has been explored in the protection of trees.

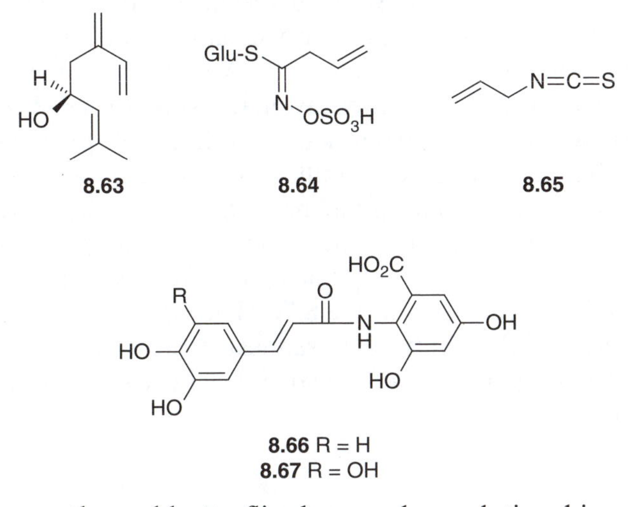

8.66 R = H
8.67 R = OH

Particular moths and butterflies have a close relationship with specific plants. The cabbage white butterfly (*Pieris brassicae*) is a garden pest on several plants of the Brassicaceae. The cabbage produces sinigrin **8.64** as a defensive substance with the enzyme β-glucosidase in adjacent tissue. When the tissue is damaged and the two are mixed, allylisothiocyanate **8.65** is produced. This deters most herbivores. However the cabbage white butterfly is attracted by sinigrin **8.64** as an oviposition stimulant. The caterpillars of the cabbage white butterfly can cope with allyl isothiocyanate and thus they develop in a non-competitive environment. Another protective feature is that the female butterfly deposits oviposition deterrents on the eggs that deter another female from laying eggs in the vicinity. Three caffeic acid derivatives, 5-deoxymiriamide **8.66**, miriamide **8.67** and its 5-glucoside, have been identified in this context. The odour of allyl isothiocyanate and its decomposition products is noticeable when cabbage is cooked or when other members of the Brassicaceae such as wallflowers are put on the compost heap. When the caterpillars of the cabbage white butterfly damage cabbage leaves, volatile semiochemicals including some terpenes such as linalool and farnesene are produced, which attract a parasitic wasp, *Cotesia rubecula*. The female of the wasp deposits her eggs within the caterpillar. These eventually develop into the larvae of the wasp and kill the caterpillar. This parasitoid and a relative, *C. glomerta*, have been used as biocontrol agents for the cabbage white butterfly.

Some of the Brassicaceae, particularly *Cheiranthus* (wallflower) species, also produce toxic cardenolides which are insect deterrents and hence they are protected. If caterpillars develop on these plants they die.

Just as plants have evolved chemical means of protecting themselves, butterflies and moths have developed the means of sequestering compounds such as these cardenolides to use as their defensive substances against predation by birds. The North American monarch butterfly (*Danaus plexippus*), which is an occasional visitor to the West Country as a result of North Atlantic gales, has been thoroughly examined in this context. The larvae which feed on milkweeds (*Asclepias* sp.) absorb toxic cardiac glycosides and some pyrazines, exemplified by **8.68** and **8.69** respectively. These are transferred to the adult butterfly which also sequesters toxic pyrrolizidine alkaloids from *Senecio* (ragwort) species. These alkaloids include retronecine **8.70**. They deter birds from feeding on the butterfly. Other butterflies sequester the bitter tasting iridoid catalpol **8.71** from plantain (*Plantago* sp.), whilst some butterflies also sequester anthocyanins.

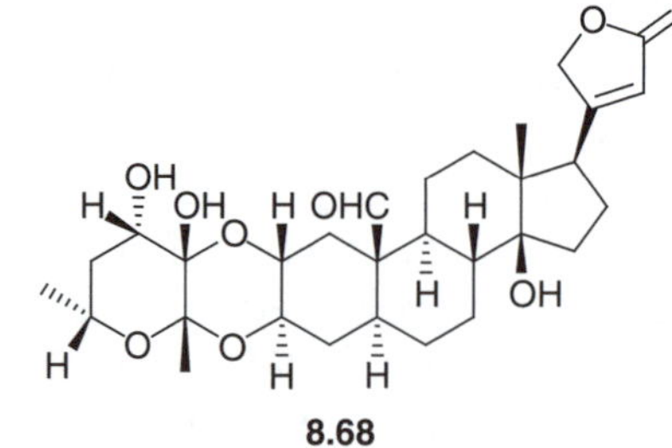

8.68

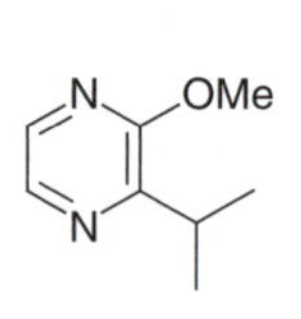

8.69

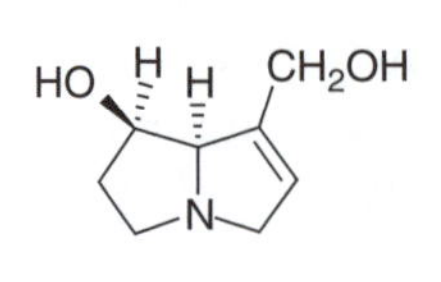

8.70

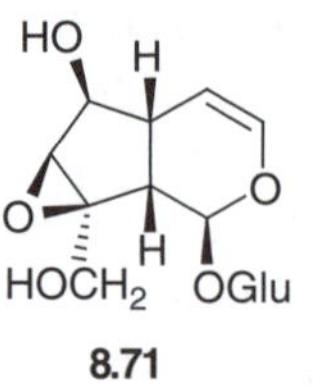

8.71

Hypericum species which include the Rose of Sharon (*Hypericum calycinum*) and particularly St. John's Wort, *H. perfoliatum*, have another way of protecting themselves against herbivores. The few insects that do feed on this plant, do so on the underside of the leaf or as leaf miners from within the leaf. The plant contains a phototoxic compound, hypericin **8.72**. On absorbing light, this compound acts as a photosensitizer for reactions involving oxygen, and it produces skin lesions around the mouth. This affects not just herbivorous insects but also livestock. Hence insects that feed on this plant have evolved ways of protecting themselves from the effects of light by feeding away from direct sunlight.

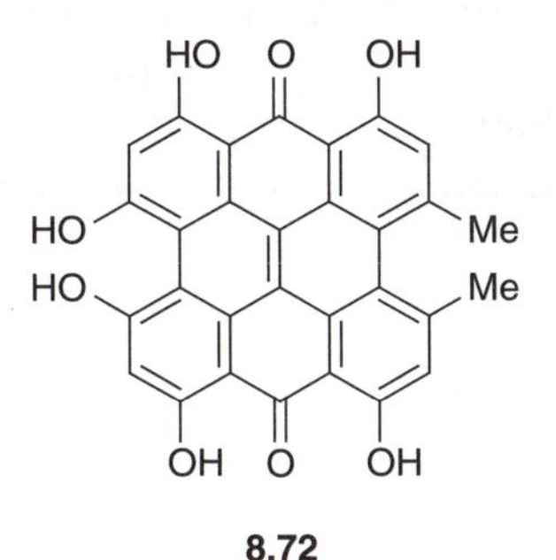

8.72

Aphids are one of the most destructive pests in the garden. Not only do they suck nutrients from the plant but they are also vectors of a number of plant viruses. Particular species are attracted to their host plant by plant volatiles. For example the sesquiterpenoid caryophyllene **8.59** is an attractant for an aphid which colonizes hops. The host selectivity of an aphid is also controlled by specific stimulants from within the plant. When an aphid lands on a plant it inserts its stylet into the plant where it comes into contact with plant products. Flavonol glycosides are possible probing cues. Other plants produce feeding deterrents such as (*E*)-2-methylbut-2-ene-1,4-diol 1-*O*-β-D-glucopyranoside **8.73**. The simple indole alkaloid gramine is a deterrent to those aphids which use barley as a host. The signalling molecule, *cis*-jasmone, may be released from the damaged plant tissue. This induces the production of volatile defensive monoterpenoids such as β-ocimene **8.55** and myrtenal from undamaged plants. These compounds not only act as aphid deterrents but also attract their predators such as ladybirds and parasitic wasps.

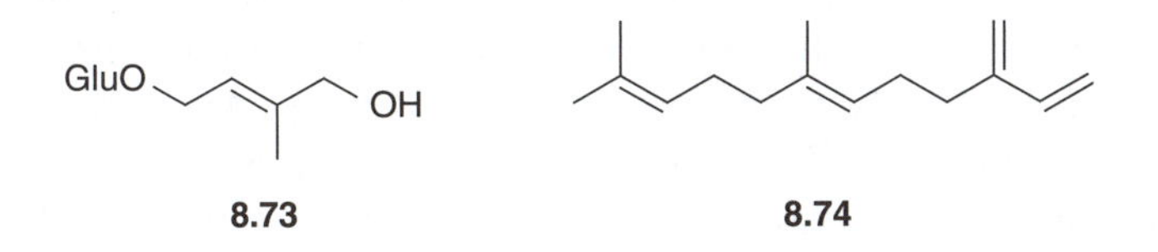

8.73 **8.74**

A number of aphid pheromones are known. Nepetalactone and its isomers are sex pheromones for some aphids. When aphids are attacked, they produce an alarm pheromone. The sesquiterpenoid (*E*)-β-farnesene **8.74** is a common alarm pheromone. This not only causes other aphids to avoid the area but it also attracts predators. In the case of the cabbage aphid, the response to the alarm pheromone is accentuated by the release of isothiocyanates from the cabbage. (*E*)-β-Farnesene **8.74** has been synthesized but it is rather unstable. However it is found in several plants including a wild potato, *Solanum berthaultii*, and in some potato cultivars. The use of the volatile essential oil from a South African wild mountain sage, *Hemizygia petiolata* (Lamiaceae), has been explored as an 'organic' method of controlling aphids. The common aphids which feed on beans, *Aphis fabae*, produce a quinone protoaphin fb **8.75** in their haemolymph, which is converted into a similar red pigment erythroaphin fb **8.76** when they are damaged by predators such as ladybirds.

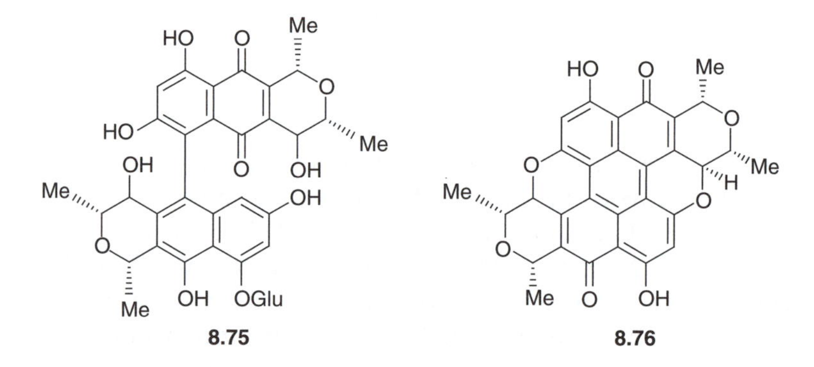

Aphids feed on the sap of a plant and exude excess sucrose as a solution, honeydew. This acts as an attractant for ants which in turn appear to protect the aphids. However the sticky surface of the plant then becomes a good medium for microbial attack.

The ladybird (*Coccinella septempunctata*) feeds on aphids. It utilizes 2-isopropyl-3-methoxypyrazine as an aggregation pheromone. The ladybird also produces a number of polycyclic quinolizine alkaloids such as coccinelline as feeding deterrents for birds.

We have already discussed the production of insect anti-feedants by plants of the Lamiaceae such as the *Salvias*. A ground-cover plant which is another member of this family and which is more commonly found growing wild in shady damp woodlands is bugle (*Ajuga reptans*). This produces a number of clerodane diterpenoids **8.77** as insect anti-feedants, allowing it to survive in a hostile environment close to the soil.

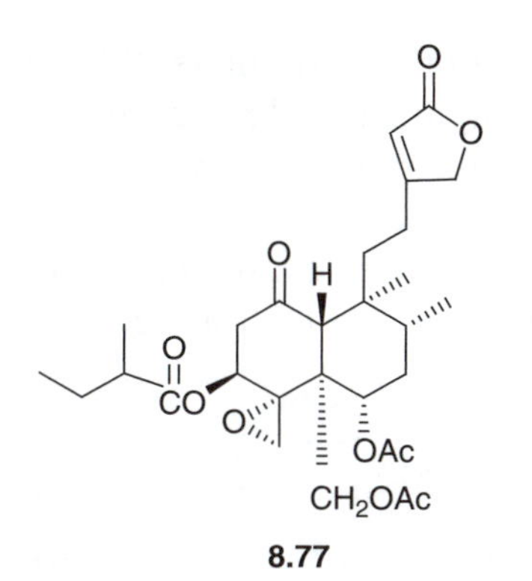

8.77

The use of pyrethrum (*Tanacetum* or *Chrysanthemum cinerariifolium*) as an insecticide has been known for centuries. The identification of the active constituents as monoterpenoid esters **8.78** led to the synthesis of a widely used family of insecticides such as permethrin **8.79**.

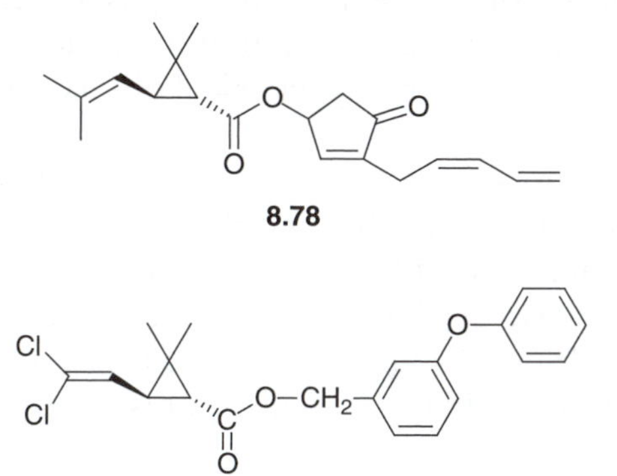

8.78

8.79

Some plants have evolved cyanogenic glucosides as a protection against herbivores such as slugs and snails. These compounds are the glucosides of α-hydroxynitriles derived from the aliphatic amino acids L-valine, L-isoleucine and L-leucine and from the aromatic amino acids L-phenylalanine and L-tyrosine. They are exemplified by linamarin **8.80**. The glycosides are relatively stable but when the plant tissue is disrupted by, for example, herbivore attack, the cyanogenic glucosides are hydrolysed by a β-glucosidase and an α-hydroxynitrile lyase. Toxic hydrogen cyanide is released, providing the plant with a defence mechanism. A common example which is found in the garden is clover, *Trifolium repens*. Clover is found in many European countries. In those countries where the average winter temperature is well below 5 °C, the plants do not produce cyanogenic glucosides, whilst in those countries where the winter temperature is above 5 °C, cyanogenic glucosides were formed. It is suggested that in the colder countries herbivores either die or hibernate during the winter and are not sufficiently prevalent early in the growing season to do significant damage to the plants. In the warmer areas where the herbivores remain active, the plants need the protection afforded by the cyanogenic glucosides during their growing period. Some butterflies can store the cyanogenic glucosides from their host plants and then use them as protection against predators.

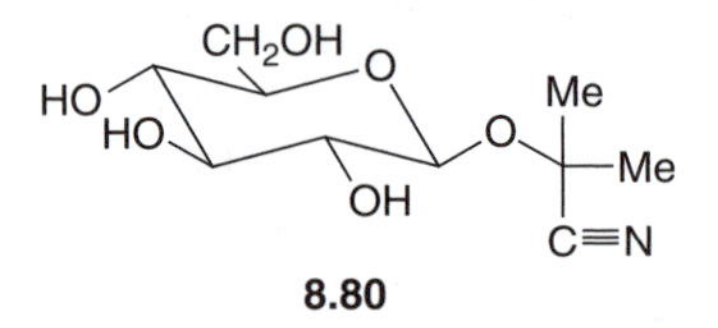

8.80

Chemistry does not just play a role in mediating interactions between plants and insects but it also determines many insect–insect interactions. Insect chemistry is a very thoroughly investigated area of chemical ecology particularly since the advent of gas chromatography–mass spectrometry and the electroantennogram. In a short book it is possible to describe only a few of the many chemical observations that have been made.

Ants are highly organized social insects. Natural products play an important part in mediating this. Trail substances marking the route between a food source and the nest are clear examples of this. The red ant, *Myrmica ruginodis*, which is found in many gardens, uses the pyrazine **8.81** as a trail substance. The pyrrole **8.82** and *Z*,*E*-α-farnesene **8.83** are also substances that are used by ants for this purpose.

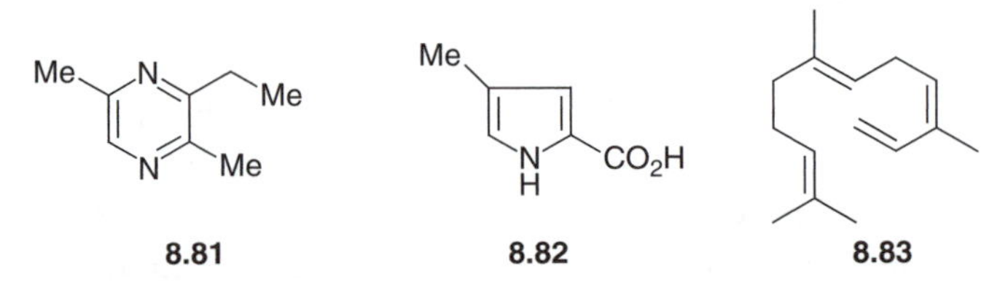

8.81 **8.82** **8.83**

The wood ant, *Formica rufa*, as a formicine ant, produces formic acid (c. 600 μg per ant) for defensive purposes. The Dufour glands of this ant also produce some undecane and tridecane hydrocarbons as well as some esters. These compounds facilitate the dispersion of the formic acid across fatty barriers on the skin of their target, increasing the effectiveness of the formic acid as a deterrent.

A number of other insects use low molecular weight acids for defensive purposes. For examples many beetles, such as carrion and dung beetles, forage in areas of high bacterial activity and yet survive. Bacteria grow best in a near neutral environment although fungi will survive in more acidic situations. Some of the carabid beetles such as

Carabus violaceus, the violet ground beetle and *C. nemoralis* produce methacrylic acid **8.84**, whilst other beetles use a surface coating of a phenol for anti-microbial purposes.

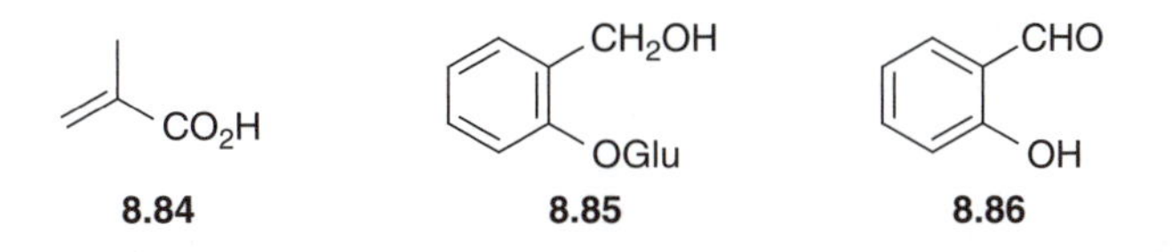

Another beetle, the Bombardier beetle, *Brachinus crepitans*, is one of a small group that has developed a chemical defence based on a quinone. When these beetles are disturbed, they squirt a hot well-aimed liquid containing benzoquinone or toluquinone at their aggressor. On their abdomen they have two glands, one containing hydrogen peroxide and phenol or cresol and the other a mixture of the enzymes catalase and peroxidase. When disturbed, they mix the enzymes and their substrate. The resultant oxidation to the quinone generates heat and enough gas to create the characteristic 'pop' and to propel the irritant liquid.

Leaf beetles use a number of chemicals to defend themselves. The larvae of *Chrysomela tremulae*, which live on willow trees, ingest salicin **8.85**. The larvae hydrolyse and oxidize this to salicylaldehyde **8.86** as a deterrent.

There are some interesting tripartite relationships involving chemical mediation. We saw an example of this with the parasitic wasp that attacks the larvae of the cabbage white butterfly. In another example, an apple emits significant amounts of *trans*, *trans*-α-farnesene **8.60** and β-ocimene **8.51** when it is infested with the larvae of the European apple sawfly. The differences in the volatiles between infested and non-infested fruit may provide a chemical clue for a parasitic wasp, *Lathrolestes ensator*, to locate the sawfly larvae in which to lay its eggs.

Spiders are very good polymer chemists. The webs that they create, whilst being very light, must not only support the weight of the spider but also absorb the kinetic energy of a flying insect without allowing it to bounce. A typical garden spider, *Araneus diadematus*, uses different silky proteins for different parts of the web. There is one for the framework and another for the spiral thread within the frame and which captures the insect. The interactions between fungi, insects and plants in the garden which have been described in this chapter reveal many examples not just of chemical warfare but also of chemical opportunism in which insects sequester and modify natural products.

Epilogue

The aim of the book has been to show that there is chemistry in every part of the garden. Even the smallest flower bed or window box can contain something of interest. The history of natural product chemistry has been a progression from isolation and structure determination through biosynthesis and function to the genetic control and regulation of biosynthesis. These developments are reflected in our understanding of the colour and fragrance of garden flowers. Had this book been written fifty years ago, it would have been concerned merely with the structures of compounds from particular plants. Today there is not only a more detailed knowledge of these constituents but also an appreciation of their function. The chemistry of the individual plant can no longer be considered in isolation but as a part of its interaction with its environment. Although we still ask the question 'What is present in a plant and how is it biosynthesized?', this is followed by 'and what is it doing here?' In this book there have been many cases of compounds that mediate interactions between, for example, plants and insects. Thus there is the tripartite relationship between the apple, the grub of the codling moth and a parasitic wasp.

A knowledge of the chemistry of these interactions can facilitate an understanding of the role of biocontrol agents. Biocontrol agents have been introduced to control greenhouse pests. On occasions the agents have been known to overstay their welcome. An understanding and use of the chemistry that mediates the interaction may help in avoiding unforeseen consequences. The use of a synthetic version of the chemical mediator affords more specific control. Since some of these substances are chemically relatively simple and work in low concentrations, there are some niche markets for the horticultural chemist.

There is a multitude of chemical problems in the garden that remain to be solved. An interesting group of problems arise from the changes to our climate. A great deal has been written on the impact of global warming and the plants that may be grown in the garden. They bring with them an interesting chemistry. However there is another side to the warmer winters. Micro-organisms and insect predators are more likely to survive the winter and hence the plant responses to these may be more pronounced earlier in the year. The correlation between the mean January temperature and the production of cyanogenic glycosides by clover may be translated into the production of similar defensive substances by other plants. Secondly a number of biennial plants require a cold period prior to flowering. If they do not experience this, flowering is poor and it may be that a gibberellin treatment might overcome this problem.

The laboratory spatula, the test tube and the TLC plate may yet have a place in the garden.

Further Reading

Books

D. H. R. Barton and K. Nakanishi (ed.), *Comprehensive Natural Product Chemistry*, Elsevier, Amsterdam, 1999.

S. Berry and S. Bradley, *Plant Life, A Gardener's Guide*, Collins and Brown, London, 1993.

J. Buckingham (ed.), *Dictionary of Natural Products*, Chapman and Hall, London, 1994.

T. P. Coultate, *Food, The Chemistry and its Component*s, The Royal Society of Chemistry, Cambridge, 4th edn, 2002.

R. J. Cremlyn, *Agrochemicals, Preparation and Mode of Action*, John Wiley, Chichester, 1991.

M. Cresser, K. Killham and T. Edwards, *Soil Chemistry and its Applications*, Cambridge University Press, Cambridge, 1993.

P. M. Dewick, *Medicinal Natural Products – A Biosynthetic Approach,* John Wiley, Chichester, 2nd edn, 2000.

J. R. Hanson, *Natural Products, The Secondary Metabolites,* The Royal Society of Chemistry, Cambridge, 2003.

J. B. Harborne, *Introduction to Ecological Biochemistry*, Academic Press, London, 4th edn, 1993.

J. B. Harborne and H. Baxter, *Chemical Dictionary of Economic Plants*, John Wiley, Chichester, 2001.

J. B. Harborne, H. Baxter and G. P Moss (ed.), *Dictionary of Plant Toxins*, John Wiley, Chichester, 1997.

D. S. Ingram, D. Vince-Prue and P. J. Gregory, *Science and the Garden*, Royal Horticultural Society and Blackwells, Oxford, 2002.

J. Mann, *Chemical Aspects of Biosynthesis*, Oxford University Press, Oxford, 1994.

J. Mann, *Murder, Magic and Medicine*, Oxford University Press, Oxford, 1992.

I. Ridge (ed.), *Plants*, The Open University and Oxford University Press, Oxford, 2002.

C. S. Sell, *A Fragrant Introduction to Terpenoid Chemistry*, The Royal Society of Chemistry, Cambridge, 2003.

R. J. Simmonds, *Chemistry of Biomolecules*, The Royal Society of Chemistry, Cambridge, 1992.

W. Steglich, B. Fugmann and S. Lang-Fugmann (ed.), *ROMPP Encyclopedia of Natural Products*, Thieme, Stuttgart, 2000.

R. H. Thomson (ed.), *The Chemistry of Natural Products*, Blackie, London, 2nd edn, 1993.

Reviews

The review journal, *Natural Product Reports*, published by The Royal Society of Chemistry, contains regular articles covering particular groups of natural products and their biological activity. Of particular interest are the series of reviews 'Recent advances in chemical ecology' by J. B. Harborne in *Nat. Prod. Rep.*, 1986, **3**, 523; 1989, **6**, 85; 1993, **10**, 327; 1997, **14**, 83 and 1999, **16**, 509. These culminated in the review '25 years of chemical ecology', *Nat. Prod. Rep.*, 2001, **18**, 361.

Specific Reviews

Chapter 1

J. R. Hanson, Development of strategies for terpene structure determination, *Nat. Prod. Rep.*, 2001, **18**, 607.

P. G. Waterman and A. I. Gray, Chemical systematics, *Nat. Prod. Rep.*, 1987, **4**, 175.

Chapter 2

R. B. Herbert, The biosynthesis of plant alkaloids and nitrogenous microbial metabolites, *Nat. Prod. Rep.*, 2001, **18**, 50.

T. Kuzuyama and H. Seto, Diversity of the biosynthesis of the isoprene units, *Nat. Prod. Rep.*, 2003, **20**, 171.

Chapter 3

A. Bajguz and A. Tretyn, Chemical characteristics and distribution of brassinosteroids in plants, *Phytochemistry*, 2003, **62**, 1027.

M. H. Beale and J. L. Ward, Jasmonates, key players in the plant defence, *Nat. Prod. Rep.*, 1998, **15**, 533.

W. Draber, K. Tietjen, J. F. Kluth and A. Trebst, Herbicides in photosynthesis research, *Angew. Chem., Int. Ed.*, 1991, **30**, 1621.

S. Fujioka and A. Sakurai, Brassinosteroids, *Nat. Prod. Rep.*, 1887, **14**, 1.

J. MacMillan, The biosynthesis of the gibberellin plant hormones, *Nat. Prod. Rep.*, 1997, **14**, 221.

L. N. Mander, Twenty years of gibberellin research, *Nat. Prod. Rep.*, 2003, **20**, 49.

T. Oritani and H. Kiyota, Biosynthesis and metabolism of abscisic acid and related compounds, *Nat. Prod. Rep.*, 2003, **20**, 414.

E. W. Weiler, Sensory principles of higher plants, *Angew. Chem., Int. Ed.*, 2003, **41**, 392.

Chapter 4

D. Staiger, Chemical strategies for iron acquisition in plants, *Angew. Chem., Int. Ed.*, 2002, **41**, 2259.

Chapter 5

C. H. Eugster and E. Marki-Fischer, Chemistry of rose pigments, *Angew. Chem., Int. Ed.*, 1991, **30**, 654.

T. Goto and T. Kondo, Structure and molecular stacking of anthocyanins – flower colour variation, *Angew. Chem., Int. Ed.*, 1991, **30**, 17.

P. Kraft, J. A. Bajgrowicz, C. Denis and G. Frater, Odds and trends, recent developments in the chemistry of odorants, *Angew. Chem., Int. Ed.*, 2000, **39**, 2980.

W. Rudiger and F. Thummier, Phytochrome, the visual pigment of plants, *Angew. Chem., Int. Ed.*, 1991, **30**, 1216.

K. Springob, T. Nakajima, M. Yamazaki and K. Saito, Recent advances in the biosynthesis of anthocyanins, *Nat. Prod. Rep.*, 2003, **20**, 288.

D. A. Whiting, Natural phenolic compounds 1900–2000, a bird's eye view of a century's chemistry, *Nat. Prod. Rep.*, 2001, **18**, 583.

Chapter 6

L. P. Christensen and J. Lam, Acetylenes and related compounds in the Astereae, *Phytochemistry*, 1991, **30**, 2453.

P. J. Houghton, Old yet new – pharmaceuticals from plants, *J. Chem. Educ.*, 2001, **78**, 175.

D. G. I. Kingston, Taxol, a molecule for all seasons, *Chem. Commun.*, 2001, 867.

D. J. Newman, G. M. Cragg and K. M. Snader, Influence of natural products upon drug discovery, *Nat. Prod. Rep.*, 2000, **17**, 215.

T. Okuda, Systematics and health effects of chemically distinct tannins in medicinal plants, *Phytochemistry*, 2005, **66**, 2012.

M. Petersen and M. J. Simmonds, Rosmarinic acid, *Phytochemistry*, 2003, **62**, 121.

H. Schildknecht, Irritant and defense substances of higher plants, a chemical herbarium, *Angew. Chem., Int. Ed.*, 1981, **20**, 164.

Chapter 7

E. Block, Organosulfur chemistry of the genus Allium, *Angew. Chem., Int. Ed.*, 1992, **31**, 1138.

T. Cornwell, W. Cohick and I. Raskin, Dietary phytoestrogens and health, *Phytochemistry*, 2004, **65**, 995.

M. H. Gordon, Dietary antioxidants in disease prevention, *Nat. Prod. Rep.*, 1996, **13**, 265.

B. Holst and G. Williamson, A critical review of the bioavailability of glucosinolates, *Nat. Prod. Rep.*, 2004, **21**, 425.

C. Jacob, A scent of therapy: pharmacological implications of natural products containing redox-active sulfur atoms, *Nat. Prod. Rep.*, 2006, **23**, 851.

H. G. Maier, Volatile flavoring substances in foodstuffs, *Angew. Chem. Int. Ed.*, 1970, **9**, 917.

M. S. C. Pedras, F. I. Okanga, I. L. Zaharia and A. G. Khan, Phytoalexins from crucifers, *Phytochemistry*, 2000, **53**, 161.

A. Stoessl, J. B. Stothers and E. W. B Ward, Sesquiterpenoid stress compounds of the Solanaceae, *Phytochemistry*, 1976, **15**, 855.

D. Strack, T. Vogt, and W. Schliemann, Recent advances in betalain research, *Phytochemistry*, 2003, **62**, 247.

Chapter 8

C. J. W. Brooks and D. G. Watson, Terpenoid phytoalexins, *Nat. Prod. Rep.*, 1991, **8**, 367.

B. Franck, Mycotoxins from mold fungi, *Angew. Chem.*, 1984, **23**, 493.

A. J. Hick, M. C. Luszniak and J. A. Pickett, Volatile isoprenoids that control insect behaviour and development, *Nat. Prod. Rep.*, 1999, **16**, 39.

M. S. C. Pedras and P. W. K. Ahiahonu, Metabolism and detoxification of phytoalexins and analogs by phytopathogenic fungi, *Phytochemistry*, 2005, **66**, 391.

F. Schroder, Induced chemical defense in plants, *Angew. Chem., Int. Ed.*, 1998, **37**, 1213.

M. Zagrobelny, S. Bak, A. V. Rasmussen, B. Jorgensen, C. M. Naumann and B. L. Muller, Cyanogenic glucosides and plant insect interactions, *Phytochemistry*, 2004, **65**, 293.

Glossary

Abscission: The loss of leaves, flowers and fruit from plants. The process is mediated by the hormone abscisic acid.
Adventitious: Shoots and roots which appear from an unusual part of a plant when, for example, a cutting is rooted.
Aglycone: The non-sugar portion, often a terpenoid, alkaloid or phenylpropanoid, of a glycoside.
Algae: Simple photosynthetic plants which are not divided into roots, stems and leaves.
Allelopathy: The release of a compound by an organism which inhibits the growth of other organisms in the locality.
Allomone: A compound, which is produced by one organism, that has a detrimental allelopathic effect on a member of another species.
Alkaloid: A large family of basic nitrogenous natural products.
Amylase: An enzyme which catalyzes the hydrolysis of starch.
Anther: The upper part of the stamen which contains the pollen.
Anthocyanin: A family of oxygen heterocycles that are biosynthesized by a combination of the phenylpropanoid and polyketide pathways and which contribute to the colours of plants.
Apical dominance: A situation in which the bud at the apex of the stem prevents the development of lateral branches. Removal of the bud allows the lateral branches to develop.
Auxin: A plant growth hormone which promotes cell elongation rather than cell division.
Bacteria: Ubiquitous microscopic unicellular organisms which are prokaryotes (lacking a true nucleus) and which have important roles in the decay of organic matter and in the fixation of nitrogen.
Betalains: A family of nitrogenous pigments which are iminium salts of betalamic acid, a tetrahydropyridine-2,6-dicarboxylic acid. They are found in beetroots.

Bioremediation: The use of biological means to restore contaminated land.
Biosynthesis: The biological synthesis of a natural product. The term biotransformation usually refers to the biological conversion of a substance that is alien to the transforming organism.
Brassinosteroids: A group of steroidal plant hormones which affect the response of a plant to stress.
C-3 and C-4 plants: Plants in which the first products of photosynthesis contain either three of four carbon atoms.
Callus tissue: Undifferentiated tissue which is often found growing over a wound.
Calyx: The collective term for the outer leaf-like part (sepal) of a flower.
Cambium: A layer of cells between the xylem and the phloem.
Carabid beetle: Ground beetles that are often brown or black and are found under wood, stones or leaf litter.
Carbohydrate: A generic name for sugars including mono-, di- and higher saccharides.
Carotenoid: A group of mainly C_{40} isoprenoid pigments with a conjugated polyene chromophore.
Cellulose: A polysaccharide composed of chains of glucose units and which forms the plant cell wall.
Chemosystematics: The use of the natural products formed by plants as part of the basis for their classification.
Chloroplast: A specialized structure within the cells of plants that contain chlorophyll and are the site of photosynthesis.
Cork: A layer of protective tissue that is found in woody plants just below the epidermis.
Cotyledon: A specialized leaf-like structure, which is often swollen with food reserves and which makes up part of the embryo in a seed. A plant which is a monocotyledon has one cotyledon and a dicotyledon has two cotyledons in the seed.
Cultivar: A variety of a plant which has been raised in cultivation rather than being obtained from the wild.
Cytokinins: A group of plant hormones which are formed in the roots and which promote cell division.
Dormancy: A resting period in plant growth and development when there is a reduced metabolic rate.
Ecdysis: The moulting or shredding of the cuticle in insects which is usually accompanied by a significant change in size or form. The process is regulated by the presence of the juvenile moulting hormone and the ecdysteroids.

Electroantennogram: When the insect antenna receptors are stimulated by interaction with a semiochemical, a potential difference is created between the tip and the base. This difference is recorded as the electroantennogram and is a measure of the receptors that are stimulated.
Endophyte: A micro-organism growing within a plant.
Etiolated: A condition in which the growing plant is weakened and is producing relatively little chlorophyll.
Feeder roots: Fine branch roots that are involved in the uptake of nutrients.
Flavonoid: Oxygen heterocycles related to the anthocyanins which form plant pigments.
Formicine ants: Ants which produce formic acid.
Fungi: Eukaryotic micro-organisms which are non-photosynthetic and which obtain their nutrients by degradation and by absorption from their surroundings.
Genus: In the taxonomic classification of plants, the level that comes between species and family. In the binomial plant name the genus comes first. Plants of the same genus have similar morphological features.
Gibberellins: A group of plant hormones that regulate several aspects of plant growth and development including germination and cell elongation.
Glycoside: A composite molecule comprising one or more sugars attached to an aglycone such as a terpenoid, alkaloid, phenylpropanoid or polyketide.
Heartwood: The wood which forms the central region of a tree trunk. It often contains resin and provides mechanical support to the tree.
Hyphae: Microscopic thread-like structures which make up the living fungus.
Hypocotyl: In a seedling it is the region of the axis between the cotyledons and the roots.
Kairomone: A group of pheromones which are interspecies specific and benefit the receiver. Thus when a pine tree is damaged by a beetle it may produce a terpene such as myrcene which lures more beetles to the tree. The myrcene is a kairomone and the beetles are the beneficiaries.
Latex: A gummy liquid exudate from a fungus or a plant.
Lichen: A composite organism which is formed by the symbiotic association between a fungus and an alga or cyanobacterium.
Lignin: A polymer formed from aromatic C_6–C_3 units which provides the strengthening material of plants.
Lipid: A hydrophobic fat-soluble natural product.
Metabolite: The generic name for compounds formed by living systems. Primary metabolites are compounds which occur in all living cells and

play a central role in the metabolism and reproduction of the cell. Secondary metabolites are characteristic of a limited range of species and many exert their biological effects on other cells or organisms.

Mitochondria: A subcellular structure (organelle) in which respiration occurs.

Mycelium: The mat of growing fungal hyphae.

Mycorrhizal fungi: Fungi which have a symbiotic relationship with the roots of a plant. They draw their nutrients from the plant whilst at the same time releasing nutrients from the soil for the benefit of the plant.

Mycotoxin: Microbial metabolites that are toxic to mammals, particularly man.

Nematode: Microscopic worms some of which parasitize insects and plants, whilst others are free-living, feeding on bacteria and fungi.

Node: The point on the stem of a plant at which one or more shoots, leaves or flowers are attached.

Nucleotide: The structural unit of a nucleic acid comprising a nucleoside (base + sugar) linked to a phosphate.

Organelle: Cellular structure which has a specialized function within the cell such as the chloroplast or mitochondria.

Oviposition: The act of depositing eggs.

Parasitoid: An organism which spends most of its life cycle within the host without eventually killing it.

Pathogen: A parasite such as a micro-organism or an insect which causes a disease within its host.

Pericarp: The wall of a ripe ovary or fruit.

Phenotype: The observable characteristics which define a particular organism.

Phenylpropanoid: A natural product which is biosynthesized by the shikimate pathway and which contains a phenyl group attached to a three-carbon chain as exemplified by cinnamic acid.

Pheromone: A substance or a mixture which is produced by a living organism that conveys a message to other members of the same species. For example there are aggregation pheromones, alarm pheromones, sex pheromones, territorial pheromones and trail pheromones.

Phloem: The cells and fibres which transport organic substances, principally sugars, from the leaves to the sites of storage such as the roots.

Photoperiodism: The response of plants to rhythmic changes in the relative length of night and day and of light intensity. A long-day plant is one that requires a long photoperiod with less than twelve hours of darkness in twenty-four hours to induce flowering (summer-flowering plants), whilst a short-day plant forms flowers when the night is longer than the day (spring- and autumn-flowering plants).

Photosynthesis: The physical process by which plants capture the energy of the sunlight and use this in a chemical reaction to convert carbon dioxide and water to sugars.

Phytoalexin: A compound produced by a plant in response to microbial attack. These are distinct from phytoanticipins which are pre-existing antimicrobial compounds.

Phytochrome: A protein–tetrapyrrole complex which absorbs red light and mediates various plant circadian and other responses.

Phytotoxin: Microbial metabolites that are toxic to plants.

Plant growth regulator: A generic term for synthetic and natural products which regulate the growth of plants in a hormonal manner.

Plastid: Major organelles found in plants which are responsible for photosynthesis (chloroplasts) and the synthesis of other compounds such as terpenes (leucoplasts), pigments (chromoplasts) and the storage of starch (amyloplasts).

Polyketide: A compound which is biosynthesized by the linear polymerization of acetate units.

Proboscis: The projecting parts of the mouth of an insect which are used in sucking in food.

Protoplast: The contents of the living cell within the cellular membrane and plasmalemma.

Rhizome: A specialized underground stem of a plant which can serve as a storage organ and as a means of spreading the plant.

Sapwood: The outer wood of a tree trunk comprising the xylem tissues and fibres involved in water transport.

Saprophyte: An organism such as a fungus which decomposes dead material externally and then absorbs the resultant nutrients.

Semiochemical: A generic term for chemicals that are involved in carrying a message. It is normally used in the context of insect chemistry to include pheromones, allomones and kairomones.

Senescence: The process of deterioration preceding the death of a plant organ such as a leaf.

Species: A basic taxonomic unit of genetically related, morphologically similar interfertile individuals. There can be subspecies and cultivars.

Spore: A small propagule of a bacterium or fungus having the function of a seed.

Stamen: The male reproductive organs of a flower.

Stylet: The primitive mouth of some nematodes and aphids which is adapted for piercing cell walls.

Symbiosis: The relationship of two organisms living together in either a mutually beneficial or parasitic relationship.
Synomone: A substance which is released by a member of one species which has a mutually beneficial effect on a member of another species.
Tachinid fly: Flies of the Tachinidae family whose larvae are internal parasites of other insects.
Tap root: The central root which forms the main axis of the root system.
Teratogen: A compound which, when ingested by a pregnant mammal, leads to deformed offspring.
Terpenoid: A natural product derived by the head-to-tail polymerization of C_5 isoprene (isopentenyl diphosphate) units. A monoterpenoid has two isoprene units, a sesquiterpenoid has three units, a diterpenoid has four units, a sesterterpenoids has five units, a triterpenoids has six units and a carotenoid has eight units.
Tuber: Swollen portions of the plant, usually the root system, which act as storage organs.
Vascular system: The combined strands of the xylem and the phloem which are responsible for the transport of water, nutrients and the products of metabolism within the plant.
Vacuole: A membrane-bound body within the cytoplasm of the cell.
Vector: An insect or animal which transmits a disease-causing organism from one plant to another.
Xylem: The part of the vascular system which is responsible for the transport of water and dissolved nutrient from the roots.

Subject Index

Abies alba, 79
abietic acid, 13, 78
abietospiran, 79
abscisic acid, 33
absinthe, 60
Acacia species, 16
acetic acid, 2,4-dichlorophenoxy (2,4-D), 31
acetophenone, 4-methoxy-, 99
acetophenone, methyltrihydroxy-, 110
acetyl coenzyme A, 9–10, 27
acetylandromedol, 114
acetylcholinesterase inhibitors, 62, 67
acetylenes, 89
  See also polyacetylenes
aconitine, 69
acoradienes, 76
Acremonium sp., 111
aeration of soils, 36
African marigold, 65
aggregation pheromones, 81, 106, 114–15
aglycones, 11, 128
agrocybin, 107
Ajuga reptans, 118
alanine, β-(3-isoxazolin-5-on-2-yl), 93
alarm pheromones, 118
alizarin, 51
alkaloids, 2, 10, 17–20, 128
allelopathic substances, 51, 63, 74, 88–9
allelopathy, 66, 93, 128
allicin, 88
alliin (S-propenylcysteine sulfoxide), 10, 88
Allium sp., 86–91
allylisothiocyanate., 90, 115
allylmethylsulfide, 88
allylpropyldisulfide, 87
aloemodin, 94
aluminium, 48
Alzheimer's disease, 62, 67
Amanita muscaria, 42, 86
Amaryllidaceae, 67
amentoflavone, 76
amino acids, 10, 63, 87–8, 93, 119
ammonia toxicity, 41
α-amylose, 22
β-amyrin, 13
anaerobic bacteria, 43
analgesics, 63, 72
anethole, 56
Anethum graveolens, 56
animal attractants, 57
animal deterents, 57
Annatto tree, 51
anthocyanidins, 16, 46–7
anthocyanins, 27, 46–50, 128
anthranilate, methyl, 53
anti-bacterial activity, 61–2, 73, 84, 88
anti-cancer activity, 68, 72–3, 84, 99
anti-fungals, 65, 76–7, 84, 88, 91
  fungal anti-fungals, 111

See also phytoalexins
anti-inflammatory activity, 88, 100
anti-microbial activity, 72, 74, 81, 100, 109
anti-oxidants, 73, 83, 97, 99–100
carotenoid and anthocyanin pigments, 46, 49
diterpenoids, 60, 62
flavonoids, 87, 91
anti-stress hormones, 33
anti-viral activity, 75
antimalarial activity, 66
antimitotics, 63, 68
Antirrhinum majus, 2, 48
antiseptics, 55
antitopoisomerase II, 73
ants, 118, 120
aphids (*Aphis* sp.), 117–18
Apiaceae, 55–6, 85
apigenin derivatives, 48–9
apiole, 57, 85
Apis mellifera, 113–14
Apium graveolens, 85
apoptosis, 100
apples, 32–3, 96, 121
Arabidopsis thaliana, 14
Araneus diadematus, 121
Armillaria mellea phytotoxins, 108
aromatherapy, 56
Artemisia sp. lactones, 66
artemisia ketone, 54
artichokes, 23
arugula, 92
Aschochyta pisi, 102–3
Asclepias sp., 116
ascorbic acid, 80, 89, 94, 98–9
Aspalathus linearis, 70
Asparagus officinalis, 88–9
asparagusic acid, 88–9
asperenyne, 89
Aspergillus sp., 7
Asteraceae, 23, 64–7
atrazine, 28
autumn crocus, 18, 68
autumnal colours, 49
auxins, 99, 128
AVG (aminoethoxyvinylglycine), 34
awobanin, malonyl-, 48
azaleas, 39
azetidine- 2-carboxylic acid, 63
Azotobacter, 39, 42
azulenes, 107

baccatin III, 10-deacetyl-, 73
Bacillus thuringiensis, 82
bark beetles, 105–6, 115
barley, 32, 117
Basidiomycetes, 106, 108
basil, 56
beans, 32, 104, 118
beeswax, 114
beetles, 105–6, 115, 120–1
beetroot, 51, 85
Begonia sp., 49
Bellis perennis, 66
bellisosides, 66
benzene, 1-allyl-4-(3-methylbut-2-enyloxy), 57
benzene, 1-allyl-2,3,4,5-tetramethoxy, 57
benzenoid aromatics, 113
benzoic acid, 3, 4-dihydroxy, 37–8
benzoic acid, 3,4,5-trihydroxy (gallic acid), 25, 47, 104
benzoquinone, 2-methoxy-6-pentyl- (primin), 71
benzylisoquinoline alkaloids, 17–18
berberine (*Berberis* sp.), 4, 18, 50–1
bergamotene, 53
Beta vulgaris, 85
betalain pigments, 51, 86, 128
betanin, 50, 85
Betula pendula terpenes, 75
Bifidobacteria sp., 23
biocontrol, 82, 122
biodegradation, 11
biodiversity of gardens, 2
bioremediation, 1, 40, 129
biosynthesis, 129
feedback regulation, 105
stages in, 11
birch, 75, 108

bitter principles, 66
bitterness, 60, 91–2, 116
bixin (*Bixa orellana*), 51
blackcurrant, 99
blue colours, 46–8
blumenol C, 89
bombardier beetle, 121
boot-lace fungus, 108
borneol, 60
botcinolides and botrydial, 104
Botrytis sp., 68, 103–5
box trees, 99
Brachinus crepitans, 121
bracket fungi, 108
Brassica sp., 32, 89–90
Brassicaceae, 11, 89–92, 115–16
brassinosteroids, 32, 129
broad beans, 104
broccoli, 89
broom, 51, 70
'brown field' sites, 1, 39–40, 42–3
bugle, 118
bulbs, bioactive constituents, 67–8
burning and germination, 83
'Burning Bush,' 57–8
bushes, transplanting, 42
but-2-ene-1,4-diol, (E)-2- methyl-, 1-O-β-D-glucopyranoside, 117
butadiene, 2-methyl (isoprene), 58
butane-2-thiol, 4-methoxy-2-methyl-, 99
2,3-butanedione, 99
butanoate esters, 96
butanoic acid, 3-methyl-, ethyl ester, 100
butanoioc acid, 2-methyl-, ethyl (+) ester, 98
2-butanone, 4-(4-hydroxyphenyl)-, 99
butein, 16
trans-3-butenoic acid, 2-amino-4-aminoethoxy- (AVG), 34
butterflies
  anti-feedants and, 5, 91, 115
  sequestration of plant toxins, 5, 116–17, 120
  sex attractants, 114
butyl acetate, 2-methyl, 96
butylisothiocyanate, 4-mercapto-, 92
butylisothiocyanate, 4-methylsulfinyl-, 90
butyric acid, α-methylene-γ-hydroxy, 68
γ-butyrolactone, α-methylene-, 68

C-3 and C-4 plants, 27–8, 129
cabbage, 91, 118
cabbage looper, 92
cabbage white butterfly, 91, 115–16
cadmium, 40
caffeic acid, 37–8, 83, 115
Calvin cycle, 27
camphor, 54–5, 60
campion, 40
capsanthin, 45
Capsicum annuum, 45, 105
capsidiol and capsinone, 105
Carabus violaceus/*C. nemoralis*, 121
caraway, 54–5
carbon dioxide fixation, 27
carboxylic acids, 8, 32
carcinogenic activity, 69
cardenolides, 116
cardiac glycosides, 62–3, 116
carnosol, 60, 62
β-carotene
  chemistry, 13, 44, 45
  occurrence, 49, 84, 89, 91
  Vitamin A and, 46
ξ-carotene, 95
carotenoids, 4, 27, 44–6, 129
carrots and carotol, 83–5
carvacrol, 55–6, 77
carvones, 54, 56
caryophyllene, 53, 81, 84, 113, 117
castalagin, 73
catalpol, 116
catechin, 87, 105
catnip or catmint, 57
cattle, 72, 106
CCC (chlorocholine chloride), 32
Ceanothus sp., 48
cedrenes and cedrol, 77

celery, 85
cellulose, 21, 25, 37, 129
Centaurea cyanus, 48
cepaenes, 87
Ceratocystis ulmi, 105–6
Cercospora rosicola, 33
chalcones, 16, 46, 75, 97
chavicols, 56–7
Cheiranthus, 116
Chelidonium majus alkaloids, 63
Chemosystematics, 20, 129
Chernobyl, 40
cherries, 51
chicory, 23, 91
chirality, 52
chlorocholine chloride (CCC), 32
chlorogenic acid, 83
chlorophylls, 26, 49
cholesterol, 82, 87
cholineesterase inhibitors, 81
chorismic acid, 15–16
Christmas rose, 63
Christmas tree, 78
chromatography, 3, 44
Chrysanthemum cinerariifolium, 119
Chrysomela tremulae, 121
chrysophanic acid, 94
Cichorium intybus and cichoriin, 91–2
1,8-cineole, 54–5, 60, 62, 99, 113
cinnamic acid, 10, 15
cinnamic acid, dihydroxy, 37–8, 83
circular dichroism, 4
citric acid, 9, 96, 99–100
citronellol, 56
Claisen rearrangement, 15
Clary Sage, 56, 60
clay soils, 36–7
clerodane diterpenoids, 60–1, 118
climate change, 6–7, 123
Clostridium perfringens, 93–4
clover, 119–20, 123
co-pigmentation, 47–8
cobalamins, 39
Coccinella septempunctata and coccinelline, 118
codling moth, 96, 122
Colchicum autumnale and colchicine, 68
Coleus sp., 50, 57
colophony, 79
Colorado beetle, 6, 81–2
colouring matters, 44–51
 See also pigments
communic acid, 77
competition inhibitors. *See* allelopathy
competitive stress, 3
Compositae. *See* Asteraceae
compost, 40–1
conifers, 75–9
coniferyl alcohol, 24, 75, 78
Consolida ajacis, 69
contact allergies, 68
contact dermatitis, 66, 71, 85
 quinones and, 50, 74
Convallaria majalis, 63
convallatoxin and convallarin, 63
coriander, 52
cornflower, 48
Cotesia sp., 116
cow parsley, 85
creosote, 25
Crocus sativus, 51
crocuses, 45
crotonic acid, 83
α-cubebene, 106
culmorin, 106
Culpepper, Nicholas, 59
cuparene, 77
Cupressus sp., 76–7
cut flowers, 31, 34, 68
cyanidin, 46–7, 84
cyanidin glycosides, 49, 91, 94, 97–9–100
cyanogenic glycosides, 119–20, 123
cyclic ethers, 52–3
cycloartenol, 82
cyclonerotriol, 106
cyclooxygenase, 87
cyclopropanecarboxylic acid, 1-amino, 33

cyclopropene, 1-methyl (MCP), 34
Cydia pomonella, 96
p-cymene, 84
cysteine, γ-glutamyl-S-1-propenyl, 87
cysteine, S-propenyl-, sufoxide
(alliin), 10, 87–8
L-cysteine, S-(2-carboxy-*n*-propyl)-, 88
Cytisus scoparius and cytisine, 51, 70
cytochrome P_{450}, 29, 99
cytokinins, 31, 129
cytotoxic activity, 25, 66–7, 75, 84, 88
cytotoxic agents, 50, 66

daffodils, 45, 67
dahlias, 64–5
daisies, 64, 66
β-damascenone, 95–6, 99–100
Danaus plexippus, 116
dandelion, 23, 66
Daphne mezereum toxins, 69
ent-dauca-5,8-diene, 76
Daucus carota constituents, 83–5
deca-4,6-diynoic acid, 66
2,4-decadienal, 93
(2E,4Z)-2,4-decadienoic acid esters, 98
defensive systems, 5–6, 34, 102, 119–21, 123
delphinidin, 16, 46–9, 99–100
Delphinium ajacis, 69
demissine, 81
5-deoxymiriamide, 115
deoxystrigol, 42
deoxyxylulose monophosphate
pathway, 14, 33, 84
depsides, 109
dermatitis
contact dermatitis, 50, 66, 71, 74, 85
furanocoumarins and, 58, 85
mezereon and, 69
polyacetylenes and, 65
detoxification, 90, 103
diallyl(poly)sulfides, 88
Dictamnus albus/gymnostylis,
dictagymin and dictamine, 57–8
digging, 1, 43
Digitalis purpurea/*D. lanata*
glycosides, 62–3
dill oil comonents, 56
diterpenoid quinones, 3, 61
dittany, white or false, 57–8
diurnal variations, 2, 52
divinorin, 60
docks, 71
dormancy, 33, 82, 129
drug bioavailability, 99
dutch elm disease, 105, 115
dyer's greenweed, 70

ecology, 4–6
eelworms, 5, 82
electroantennograms, 120, 130
elemicin, 57
ellagic acid and ellagitannins, 25, 73
emodin, 94
endophytic organisms, 111, 130
5-enolpyruvyl-shikimate-3-phosphate
synthase (EPSP), 16
enzyme inhibitors, 99–100
enzymes
extracellular, 103–4, 112
polyphenol oxidase, 83
transfer between species, 103
ergosterol, 15, 29
ergovaline, 111
ericaceous mycorrhizal fungi, 42
Eruca sativa, 92
erythroaphin, 118
erythrose 4-phosphate, 15
Escherichia coli, 94
estragole, 56
ethylene, 33–4, 96
eugenol, 56, 113
Euphorbia sp., 69

fairy rings, 42, 107–8
falcarinol and falcarindiol, 65, 84
Fallopia japonica, 6
farnesenes, 113, 116, 118, 120–1
farnesol, 53
fatty acids, 10, 12

fenchone, 56
fennel, 56
fertilisers, 40–1
'fescue foot,' 106
feverfew, 66
fires, 83
flatulence, 23, 93
flavans, hydroxy-, 68
flavonoids, 30, 75, 89, 130
  See also catechin; keampferol;
    myrcetin; quercetin
flavoxanthin, 45
flowers
  cut flowers, 31, 34, 68
flowering out of season, 32
  scent of, 52–3
fly agaric, 42, 86
foaming response, 13
Foeniculum vulgare, 56
food colourings, 51
Formica rufa and formic acid, 120
forskolin, 50
foxgloves, 62–3
Fragaria x ananassa, 98
fragilin, 72
free radicals, 24–5
  See also radical scavengers
fructose, 9, 23, 27
fruit
  ripening, 2, 33–4, 95–6
  soft fruit, 98–101
  storage and transit, 34, 112
fruit trees, 96–8
fulvic acids, 37
fungi, 130
  apples and, 97
  climate change and, 7
  interactions between, 111–12
  interactions with plants, 102–9
  phytoalexin response, 93
  phytotoxin production, 5
  potatoes and, 81, 83, 105
  tomatoes and, 94–5
  trace metals and, 40
  See also endophytic; lichens;
    mycorrhizal
fungicides, 29
2(H)-furo-[2,3]-pyran-2-one, 3-
    methyl-, 83
Fusarium sp., 68, 94–5, 105–6

α-galactosidase, 93
galanthamine, 18, 67
gallic acid, 25, 47, 104
garlic, 10, 88
gas chromatography/mass
    spectrometry, 30, 52, 120
genistein and *Genista tinctoria*, 70
geosmin, 41, 43, 86, 112
geraniol, 10, 13, 52–3, 99–100, 113
geraniums, 49
geranyl acetate, 66
germination inhibitors, 83
giant hogweed, 6, 85
Gibberella pulicaria, 105
gibberellic acid, 32, 82
gibberellins, 82, 99, 123, 130
  biosynthesis, 29–30
  structure elucidation, 4, 31–2
gladioli, 4, 12
gliotoxin and *Gliocladium roseum*,
    111–12
global warming, 6–7, 123
glucobrassicin, 91
glucoerucin and gluconasturtiin, 92
glucoraphanin, 90
glucosinolates, 11, 89–92
glucuronic acids, 23
glyceraldehyde-3-phosphate, 27
glycerol, 1,3-O-di-*p*-coumaroyl-, 89
glycine, aminoethoxyvinyl (AVG), 34
glycoproteins, cell wall, 22
glycosides, 11, 130
glyphosate, 16
goitrin, 91
goldenrod, 51
gout, 60, 68
gramine, 117
grapes, 100, 103–5
grass, 42, 54, 65
  See also lawns
grayantoxins, 114

greater celandine, 63
growth hormones, 30
growth inhibitors. *See* allelopathy
growth regulators, 32, 132
gums, 23

haem complexes, 24, 29, 40, 99
hallucinogens, 60–1
harlequin ladybird, 6
harzianopyridone, 112
heart effects, 62–3, 70, 100
Hedera helix and hederacoside, 71
Helianthus annuus, 32
Helicobacter pylori, 88, 90
Helleborus niger, 63
Hemizygia petiolata, 118
hepta-1,3-diyne-5-ene, 1-phenyl, 65
1,9-heptadecadiene-4,6-diyne-3-ol (falcarinol), 65, 84
3-heptanol, 4-methyl-, 106
heptatriyne, phenyl-, 65
heptulosonic acid, 15
herbal teas, 56, 60, 70
herbalists, 59–60, 62
herbicides, 16, 28, 31, 65
herbivores, 34, 73, 119–20
herbs, 54–7
Heterodera rostochiensis, 82
cis-hex-3-enol, 66
2-hexanal and 2-hexen-1-ol, 96
3-hexen-1-ol and *trans*-2-hexenal, 54
Z-3-hexen-1-ol, 99
Hill reaction, 27–8
hinokiflavone, 76
Hippocrates, 72
HIV virus, 75
homoserine, 93
honey bee, 113–14
honey fungus, 108
honeydew, 118
hops, 117
hormones, 30–4
house fly, 115
humic acids, 37–8
humulene, 84
Hydrangea macrophylla, 1, 48
hydrogen bonding, 21
hydroxystilbenes, 78–9, 100
hypericin, 117
Hypericum sp., 117

indigo, 51
indolyl-3-acetic acid, 30, 111
infrared spectroscopy, 4
inoleic acid, 10
insect anti-feedants, 65, 73, 81, 91–2, 117
  diterpenoids from *Salvia* sp., 3, 61, 118
  seasonal variation, 5–6
  See also pests
insect attractants, 52, 96, 105, 113–14
  See also pollination
insect repellents, 55, 96, 115
insect vectors, 105
insecticide resistance, 82
insecticides, 55, 65, 70, 119
inulin, 23, 66
ionones (α- and β-), 84, 95, 99
Ips sp. and ipsdienol, 115
iron, 29–30, 39–40, 42, 83
irritants. *See* dermatitis
isatin B and *Isatis tinctoria*, 51
isoborneol, methyl-, 112
isoflavones, 93
isomerism, geometrical, 52
isopentenyl (3-methylbut-3-en-1-yl) pyrophosphate, 10, 14
isopentyl isovalerate, 95
isoprene, 57–8
isoprenoid C_5 units, 10–11, 13, 84
isopulegol, 55
isoquinolines, benzyl-, 63–4
isothiocyanates, 90, 92, 115, 118
isotope ratio, sucrose, 28
isotrichodermin, 106
isovelleral, 108
isowillardine, 93
Italian cypress, 77
ivy, 71

Japanese knotweed, 6

Japanese rose, 53
Jasmin (*Jasminum officinale*), 53
cis-jasmone, 53, 117
jasmonic acid, 34, 53, 82
juglone, 74
Juniperus sp., 77

kaempferol (glycosides), 47–8, 87, 99
kairomones, 5, 130
ent-kaurene, 29
kinetin, 31

Labiatae. *See* Lamiaceae
Laburnum anagyroides, 69
laccase, 104
lachrymators, 87
Lactarius sp. azulenes, 107
Lactuca sp., lactucin and
lactucopicrin, 91–2
ladybirds, 6, 117–18
Lamiaceae, 55, 57, 59–62, 118
lanosterol, 15
larch (*Larix deciduas*) and larixol, 78
larkspur, 69
Lathrolestes ensator, 121
Lathyrus odoratus, 53
laudanosine, 64
lavender (*Lavandula angustifolia*), 54
lawns, 2, 26, 36, 106
leaf exudates, 92
leaf scents, 54–8
Lecanora sp. and lecanoric acid, 109
legumes, 93–4, 110
lemon balm, 56
Leptinotarsa decemlineata, 6, 81–2
lettuce, 3, 32, 91–2, 103
seedlings and allelopathics, 89, 93
lettucenin A, 92
leucocyanidin, 97
leukaemia, 66, 72
Leyland cypress, 76
lichens, 109–10, 130
lichesterinic acid and lichexanthone,
109
lignans and lignases, 25
lignin, 24, 37, 40, 78, 130
lilacs, 48, 53
Liliaceae, 67, 88
lily-of-the-valley, 6, 63
limonene, 56, 65, 79, 85
linalool, 52–4, 99–100, 113, 116
linalyl acetate, 56
linoleic acid, 96
linolenic acid, 54
5-lipoxygenase, 100
liver, 90, 99
livestock toxicity, 106, 111, 117
lobsters, 46
lolines and *Lolium perenne,* 111
lubimin, 105
lup-20(29)-ene-3β,28-diol, 75
lupeol, 91
lupin (*Lupinus polyphyllus*) alkaloids,
69–70
lutein, 45 49, 65, 91
lycoctonine, 69
lycopene (from *Lycopersicon
esculentum*), 45, 94–6
lycorine, 18, 67
β-lycotetraose, 94

maackain, (+)-6a-hydroxy, 103
madder, 51
magnesium deficiency, 40
malic acid, 9, 96, 100
Malus domestica, 96
malvidin, 100
manool, 78
manuring, 41
Marasmius oreades and marasmone,
42, 107–8
marigolds, 45, 65
marjoram, 56
mass spectrometry, 4, 19
See also gas chromatography
cis-matricaria methyl ester, 64
MCPA (2-methyl-4-
chlorophenoxyacetic acid), 31
meadow sweet, 72
Melissa officinalis, 56
mellein, 5- and 6-methoxy-, 84–5,
106

Mentha sp., menthol and its derivatives, 13, 54–56
metal ions, 1, 36–9, 48
methacrylic acid, 121
methionine, 12
mevalonate pathway, 14, 33, 84
microbes within the soil, 41–3
microwaves, 90
migraine, 66
milk caps, 107
milkweeds, 116
minerals in garden soils, 35–7
mints and mintlactone, 13, 54–5
miriamide, 115
mistletoe, 75
monarch butterfly, 116
Monilinia fruticola, 93
monkshood, 69
monoterpenoids, 53, 113
morphine family, 18, 60, 63–4
morpholine fungicides, 15
moths, 2, 52, 96, , 114–5
δ-multistriatin, 106
Musca domestica and muscalure, 115
mustard oils, 90
mutagens, 97
mycorrhizal fungi, 5, 41–2, 110, 131
mycotoxins, 106, 131
Myiopharus doryphorae, 82
myrcene, 66, 84, 113, 115, 130
myrcetin, 87
myristicin, 57, 85
Myrmica ruginodis, 120
myrosinase, 90–2
myrtenal, 117

NADP, 27
naphthalene, 1,4,5-trihydroxy-, 74
narciclasine (*Narcissus pseudonarcissus*), 67–8
natural product families, 8
nematodes, 65, 131
See also eelworms
neomenthol, 55
Neotyphodium sp., 111
nepetalactone (*Nepeta cataria*), 57, 118
nerol, 52–3, 113
neurotoxins, 60
neurotransmitters, 19–20, 71, 75
Nicotiana sp., 70, 103
nicotine, 10, 17
'night-scented' stocks, 2, 52
nitrogen deficiency, 40–1
nitrogen fixing bacteria, 39, 42, 110
nivalenol, 106
NMR spectrosopy, 3–4, 19, 93
nonacosane, 89
2,6-nonadienal, 93
Norwegian spruce, 78
nuclear magnetic resonance (NMR), 3–4, 19, 93
nuclear Overhauser effect, 4
nutrients, soil, 38–41

oak, 73
β-ocimene, 53–4, 113, 117, 121
Ocimum basilicum, 56
R-(+)-octane-1,3-diol, 96
6-octen-2-one, (S)-1,3-dihydroxy-3,7-dimethyl-, 81
odour. *See* scents
oil of wintergreen, 72
oil seed rape, 32
oleanolic acid, 75
olivine, 37
onion, 10, 87, 104
orange pigments, 44
orchids, 42
'organic' agriculture, 6, 80, 119
organic content of soils, 37–8
Origanum vulgare, 56
ornithine, 10, 17
orsellinic acid, 109
oxalic acid, 94
oxidative coenzymes, 29–30
trans-9-oxodec-2-enoic acid, 114

paclitaxel, 72, 111
paclobutrazole, 32
pain killers, 63, 72
palmitic acid, 114
pansies, 45

Papaver sp., 18, 63
papaverine, 18
paraquat, 28
parasitic wasps, 116–17, 121, 122
parietinic acid, 110
parsley, 56
parsnips, 85
parthenolide, 66
particle size, soils, 35
Pastinaca sativa, 85
pathogens, 103, 131
  See also fungi
patulin, 97
pears, 98
peas, 32, 93, 102
pectic acids and pectins, 23
pelargonidin and its glycosides, 46–9, 98
Penicillium sp., 4, 7, 12, 97
pennyroyal, 55
pentan-2-one, 4-mercapto-4- methyl-, 99
peonidin, 46, 100
peppers, 45, 105
peramine, 111
perfumery, 60
permethrin, 119
Peronospora parasitica, 91
pests, 55, 96, 114, 122
  See also Colorado beetle
Petroselinum crispum, 56
pH, 39–40, 46
Phaseolus vulgaris, 32–3
β-phellandrene, 56
phenethylisothiocyanate, 92
phenol coupling reactions, 24–5, 76, 110
  in alkaloid biosynthesis, 18, 64, 67
phenoxyacetic acid, 2-methyl-4-chloro- (MCPA), 31
phenylpropanoids, 15–17, 56, 113, 131
  biosynthesis from, 11
polyketide combination with, 16–17
  as secondary metabolites, 2
phenylpyruvic acid, 15
pheromones, 5, 114, 118, 131
phloretin and phloridzin, 97
phloroacetophenone, methyl-, 110
Phomopsis oblonga, 106
phosphate availability, 38
phosphoenol pyruvate, 9
phosphokinase C, 75
phosphorus deficiency, 40
photoperiodism, 28, 82, 131–2
photosensitivity, 117
photosynthesis, 26–8, 132
phthalide, 3-butyl-, and derivatives, 85
physcion, 110
phytoalexins, 5, 83, 92–3, 100, 104
  interspecies transfer, 103
  phytoanticipins and, 102, 132
  resveratrol, 100, 105
phytoanticipins, 102, 132
phytochelatins, 40
phytochrome, 28, 132
phytoene and phytofluene, 95
Phytophthera infestans, 81, 83, 102
phytotoxins, 31, 103–9, 132
Picea abies and piceatannol, 78
Pieris brassicae, 95, 115–16
pigments, 50–2, 86, 95, 99
  See also colouring matters
pimaric acid, 77–8
Pinaceae, 75
α-pinene, 79
  in herbs, 55–6, 60, 62
  as an insect attractant, 113, 115
β-pinene, 55–6, 79
pinoresinol, 25
pinosylvin, 78
Pinus sylvestris, 79
piperitone, 65
Piptoporus betulinus, 108–9
pisatin, 93, 102–3
Pisum sativum, 32, 102
plant classification by chemosystematics, 20
plant hormones, 30–4
plantain (*Plantago* sp.), 116
platelet aggregation, 87

Pliny (Gaius Plinius Secundus), 72
pollination and colour, 49
pollination and scent, 2, 52, 113–14
pollution and lichens, 109
polyacetylenes, 64–6, 84, 107
polygalactouronic acids, 23
polygalaic acid, 66
Polygonum cuspidatum, 6
polyketides, 12, 132
  biosynthesis, 10, 14, 27
  lichens, 109–10
phenylpropanoid combination with, 16–17
  as secondary metabolites, 2
polyphenol oxidase enzymes, 83
Polyporus betulinus and polyporenic acid, 108–9
polysaccharides, 22–3, 38
poplar and populin, 72
poppies, 18, 63–4
porphyrin derivatives, 26, 29
potassium deficiency, 40
potato, 80–3
  Colorado beetle and, 6, 81–2
  farnesene in, 118
  marigolds and, 65–6
  phytoalexin discovery, 102
  tuber formation, 34
pregnanes, 62
prephenic acid, 15
primary metabolites, 2, 130
primin, 71
primrose (*Primula* sp.), 49, 71
proanthocyanidins, 97
progoitrin, 91
prop-2-enethioate, S-methyl, 89
prostaglandins and prostacyclins, 87
protoaphin, 118
protocyanin, 48
pruning, 103
Pseudomonas aureofaciens/P. putida, 42
Psilolechia lucida, 109
psoralen, 58, 85
psychotropic agents, 20, 60
pterocarpans, 93
pulegone, 55
purgatives, 94
pyran polyacetylenes, 65
pyrazine, 2-alkyl-3-methoxy-, 93, 116, 118
pyrazine trail pheromones, 120
pyrethrum, 119
pyrone, 6-pentyl-, 112
pyrroles, 120
pyrrolizidine alkaloids, 116
Pyrus communis, 98
Pythium sp., 65, 112

queen bee substance, 114
quercetin and its glycosides, 47, 49–50, 87, 91
Quercus robur, 73
quinic acid derivatives, 48, 83
quinolizidine alkaloids, 69–70
quinones, 51, 121

radical scavengers, 49, 89, 97
radiolabelling studies, 12, 19
radish, 33, 90
raffinose, 93
ragwort, 116
Raphanus sativus, 33, 90
raspberries, 99
red pigments, 44
redcurrant, 99
resveratrol, 78, 100, 103, 105
reticuline, 18, 64
retronecine, 116
rhein and *Rheum rhaponticum*, 94
Rhizobium bacteria, 22, 39, 42
Rhizoctonia solani, 112
rhododendrons, 6, 39, 42, 114
rhoeadine alkaloids, 63
rhubarb, 94
Ribes nigra, 99
ribulose 1,5-diphosphate, 27
rishitin, 83, 105
rocket lettuce, 92
rocks, contributing to soils, 36
root exudates, 5–6, 38, 93
root vegetables, 80–6

rooting hormones, 30
roots and polysaccharides, 38
Rorippa nasturtium - aquaticum, 92
Rosa rugosa, 53
rose, 3, 46–7, 52–3
See also geraniol
Rose of Sharon, 117
rosemary (*Rosmarinus officinalis*), 55, 62
rosenonolactone, 97
rosmanol and rosmarinic acid, 62
Rubia tinctorum, 51
Rubus ideaus, 99
Rumex sp., 71
rye grass, 111

sabinene, 56, 84
saffron, 51
sage (*Hemizygia petiolata*), 118
sage (*Salvia officinalis*), 56, 59–60
St. John's Wort, 117
Salix sp. and salicylates, 72, 121
salutaridinone, 18
Salvia sp., 3, 48–9, 59–61
sanguinarine, 63
Santolina chamaecyparis, 54
santonin, 13, 66
saponins, 13, 66, 71, 89, 94
Sarothamnus scoparius, 70
sarsasapogenin, 89
sawfly, 121
scents, 2, 52–8, 99
sclareol and sclareolide, 60
Sclerotinia sp., 97
Scolytus multistriatus, 105–6, 115
scoparin, 70
scopolin, 83
Scots pine, 79
sea wormwood, 66
seasonal variations, 5, 59–61
secondary metabolites, 2, 8, 11, 131
sedanenolide, 85
seed germination, 34, 83
selenium, 88
β-selinene, 85
semiochemicals, 5, 7, 114, 116, 132
Senecio sp., 116
sequestration of toxins, 5, 116–17, 120
Seriphidium maritimum, 66
sesquiterpenoid lactones, 3, 66, 91–2
sesquiterpenoids, 13, 76, 81, 104, 107–8
sewage sludge, 40
sex attractants, 114, 118
sheathing mycorrhizal fungi, 41–2
shikimic acid pathway, 15–16
Silene vulgaris, 40
silica, 36
silver fir, 79
sinapyl alcohol, 24, 75
sinigrin, 91–2, 115
β-sitosterol, 91
slugs and snails, 119
snapdragon (antirrhinum), 2, 48
'snow fungus', 106
soft fruit, 98–101
soils
contamination, 39–40
freshly dug, 1, 41
microbes, 41–3
mineral structure, 35–7
nutrients from, 38–9, 40–1
organic content, 37–8
pH effects, 39–40
waterlogging, 43
Solanaceae, 6, 81, 105
See also potato; tomato
solanine and solasodine,, 81
solanoeclepin A, 82
Solanum sp., 81, 118
See also potato
Solidago canadensis, 51
sparteine, 70
spider mite, 55
spiders, 121
splendidin, 61
starch, 22–3, 32, 80
stavesacre oil, 69
steroids, 10, 14, 32
stilbene oxidase, 105
stilbene synthase, 103

stinging nettle, 71
stomach ulcers, 72, 88, 90
Stone, Rev. Edward, 72
strawberries, 98
Streptomycete sp., 43, 112
strigol, 110
strophanthidin, 63
structural materials, 2, 21–5
sucrose, 28, 86
sugars, 9, 11
sugiol, 77
sulforane, 90
sulfur, 86
sunflower, 32
sweet pea, 2, 53
symbiosis, 109, 111, 133

tachinid fly, 82, 133
Tagetes sp., 45, 65
Tanacetum sp., 66, 119
tannins, 25
tanshinones, 61
tansy, 66
taraxanthin, 45
taraxacin (*Taraxecum* sp.), 66
tartaric acid, 100
taxine, 72
taxol (paclitaxel), 72, 111
Taxus sp., 72–3, 111
teratogenic effects, 70, 81, 133
terpenoids, 2, 13–15, 133
  biosynthesis from, 11
  biosynthesis of, 10
  carotenoids as, 13, 44
1-terpinen-4-ol, 56
terpinolene, 65
tetradecanal, 53
Tetranyctus urticae, 55
Thale Cress, 14
thiopropanal-S-oxide, 87
thujaplicins, 77
thujone, 60
thymol, 55–6
tobacco, 70, 103
α-tocopherol, 89
α-tomatine, 94
tomato, 31, 45, 94–6
toxic metals, 1, 40
toxicity, 81, 91, 94
toxins
  from ornamental plants, 69–71
  sequestration by butterflies, 5, 116–17, 120
  See also phytotoxins
trail pheromones, 120
transplanting, 42, 110
transport of foodstuffs, 6, 112
trees
  conifers, 75–9
fruit, 96–8
  ornamental, 72–5
tremuloidan, 72
triadimefon, 29
tricarboxylic acid cycle, 9, 109
Trichoderma sp., 111–12
Trichoplusia ni, 92
Trichothecenes (*Trichothecium roseum*), 97, 106
Z-9-tricosene, 115
Trifolium repens, 119
tripartite relationships, 5, 121, 122
truffles, 42
tuberonic acid, 82
tulips (*Tulipa* sp.), 68, 104
turpentine, 79

ultraviolet screening, 49
ultraviolet spectroscopy, 3, 64–5
urine, 89
Urtica dioica, 71
usnic acid (*Usnea* sp.), 110

vegetables
  colour loss on boiling, 26–7
  root vegetables, 80–6
velleral, 108
velutinal, 107
verbenone, 115
vescalene, 73
vesicular-arbuscular fungi, 42
viniferin, 100
violaxanthin, 45

viridin, 111
viscatoxins, 75
Viscum album, 75
Vitamin A, 46
Vitamin C, 80, 89, 94, 98–9
Vitis vinifera, 100, 103
vitispirane, 100
vomitoxin, 106
vulgaxanthin, 86

wallflower, 116
walnut, 74
water absorption by cellulose, 22
water cress, 92
water stress, 33
wax coatings, 22
'white-rot' fungi, 25
willardine, 93
willow, 72, 121
wine, 73, 100, 103–4
winter temperatures, 119–20, 123
woad, 51
wood-rotting fungi, 25
wormwood, 13
wyerone acid, 104

X-ray crystallography, 4, 48, 82
xanthoperol, 77
xanthorin (*Xanthoria* sp.), 110

yellow pigments, 44, 51
yew, 72

zeaxanthin, 45